Student Edition

Eureka Math
Geometry
Modules 3, 4, & 5

Special thanks go to the Gordon A. Cain Center and to the Department of Mathematics at Louisiana State University for their support in the development of *Eureka Math*.

For a free *Eureka Math* Teacher Resource Pack, Parent Tip Sheets, and more please visit www.Eureka.tools

Lesson 1: What Is Area?

Exploratory Challenge 1

a. What is area?

b. What is the area of the rectangle below whose side lengths measure 3 units by 5 units?

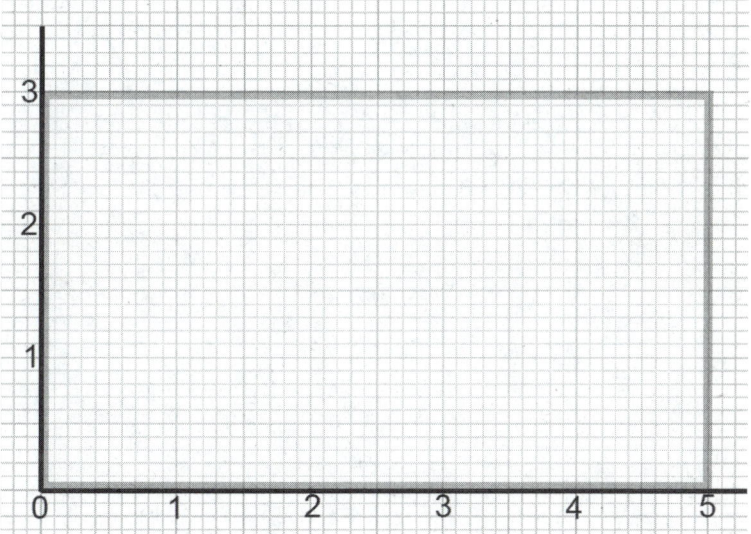

c. What is the area of the $\frac{3}{4} \times \frac{5}{3}$ rectangle below?

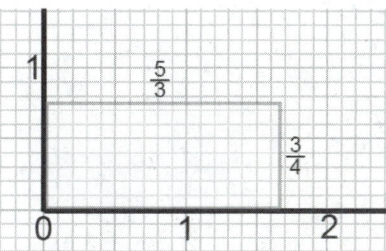

EUREKA MATH™

© 2015 Great Minds. eureka-math.org
GEO-M3-SE-B2-1.3.0-10.2015

Exploratory Challenge 2

a.   What is the area of the rectangle below whose side lengths measure $\sqrt{3}$ units by $\sqrt{2}$ units? Use the unit squares on the graph to guide your approximation. Explain how you determined your answer.

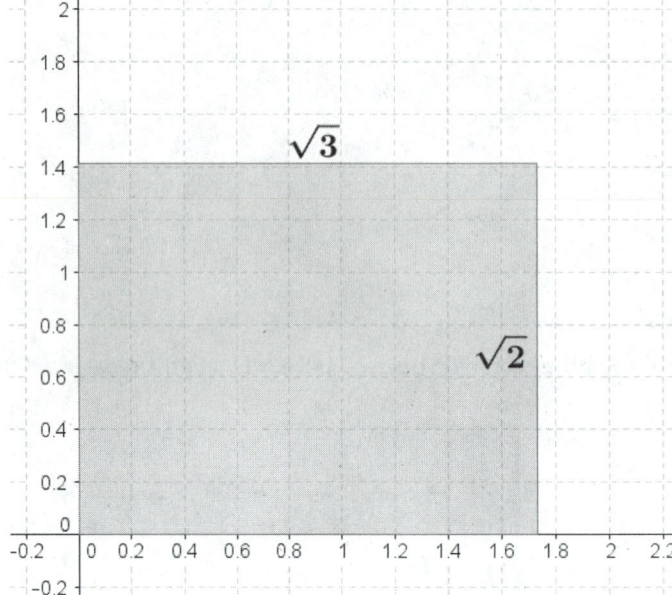

b.   Is your answer precise?

Lesson 1:   What Is Area?

EUREKA
MATH™

Discussion

Use Figures 1, 2, and 3 to find upper and lower approximations of the given rectangle.

Figure 1

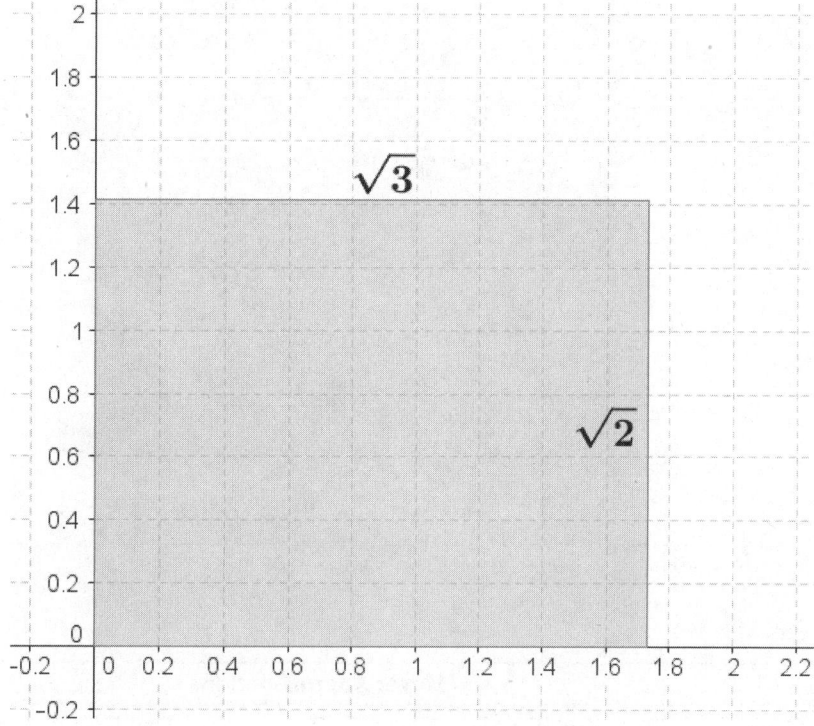

Figure 2

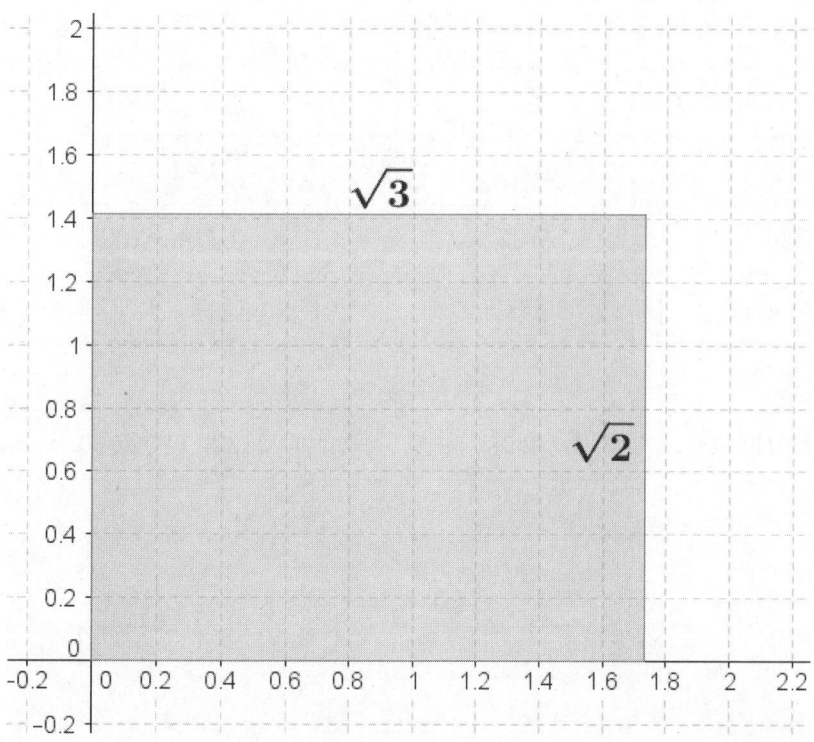

EUREKA
MATH™

Figure 3

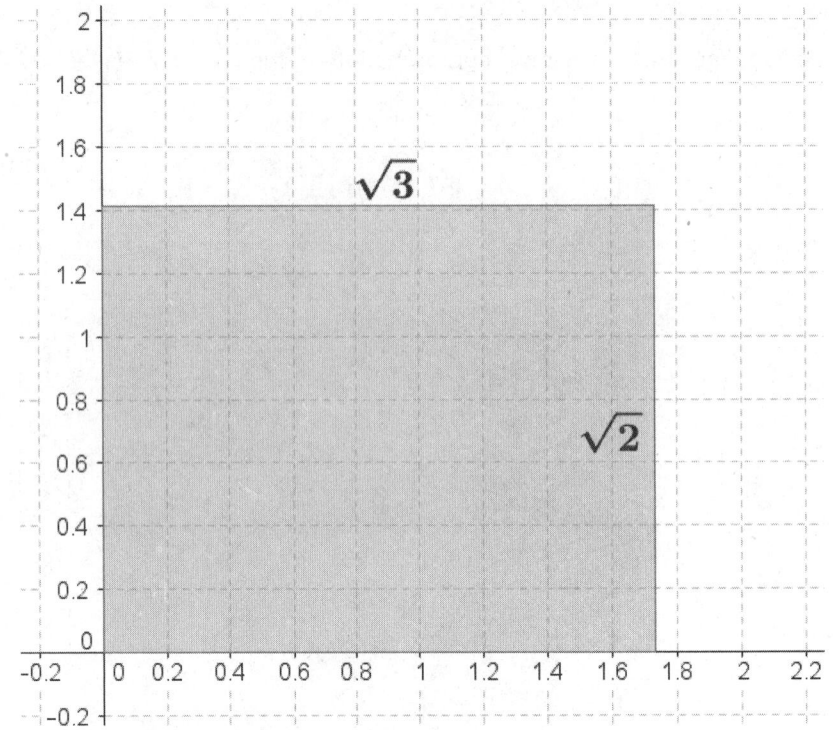

Lower Approximations		
Less than $\sqrt{2}$	Less than $\sqrt{3}$	Less than or equal to A
1	1	$1 \times 1 =$
	1.7	$\times 1.7 =$
1.41		$1.41 \times \quad =$
1.414	1.732	$1.414 \times 1.732 =$
1.4142	1.7320	$1.4142 \times 1.7320 = 2.449\,344$
		$= 2.449\,482\,430\,5$
1.414\,213	1.732\,050	$1.414\,213 \times 1.732\,050 = 2.449\,487\,626\,65$

© 2015 Great Minds. eureka-math.org
GEO-M3-SE-B2-1.3.0-10.2015

Upper Approximations		
Greater than $\sqrt{2}$	Greater than $\sqrt{3}$	Greater than or equal to A
2	2	$2 \times 2 = 4$
1.5	1.8	$1.5 \times 1.8 =$
1.42	1.74	$1.42 \times 1.74 = 2.4708$
	1.733	$\times 1.733 =$
1.4143	1.7321	$1.4143 \times 1.7321 = 2.449\,709\,03$
1.41422	1.73206	$1.41422 \times 1.73206 = 2.449\,513\,893\,2$
		$= 2.449\,490\,772\,914$

Discussion

If it takes one can of paint to cover a unit square in the coordinate plane, how many cans of paint are needed to paint the region within the curved figure?

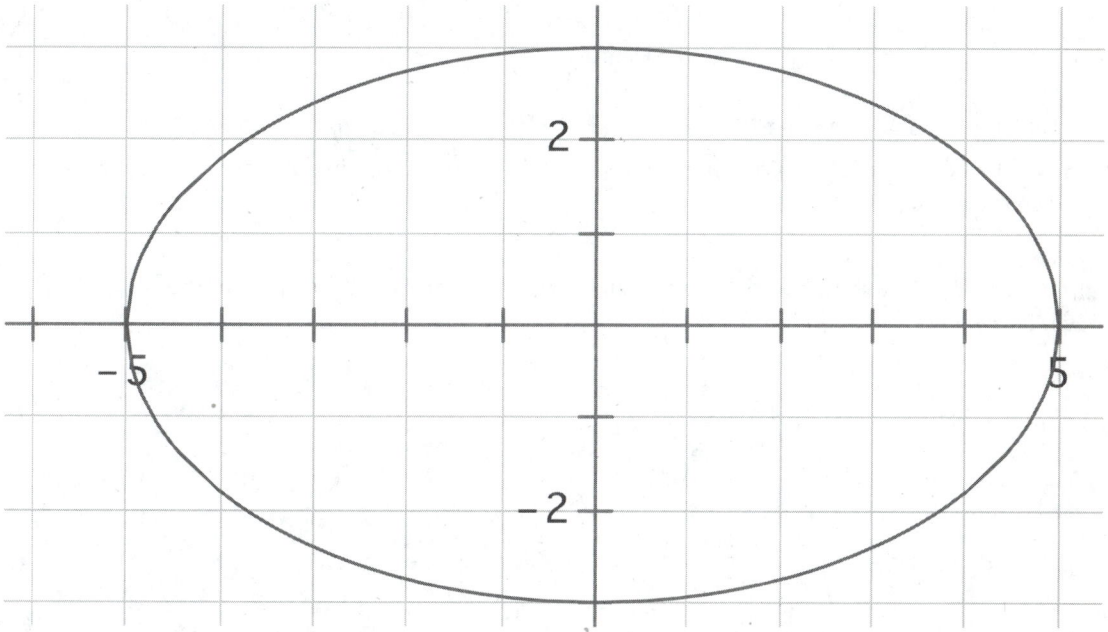

EUREKA
MATH™   Lesson 1:   What Is Area?   S.5

          © 2015 Great Minds. eureka-math.org
          GEO-M3-SE-B2-1.3.0-10.2015

Problem Set

1. Use the following picture to explain why $\dfrac{15}{12}$ is the same as $1\dfrac{1}{4}$.

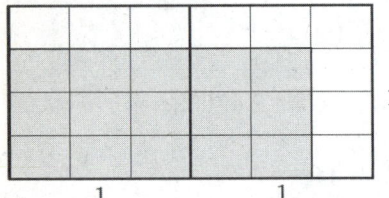

2. Figures 1 and 2 below show two polygonal regions used to approximate the area of the region inside an ellipse and above the x-axis.

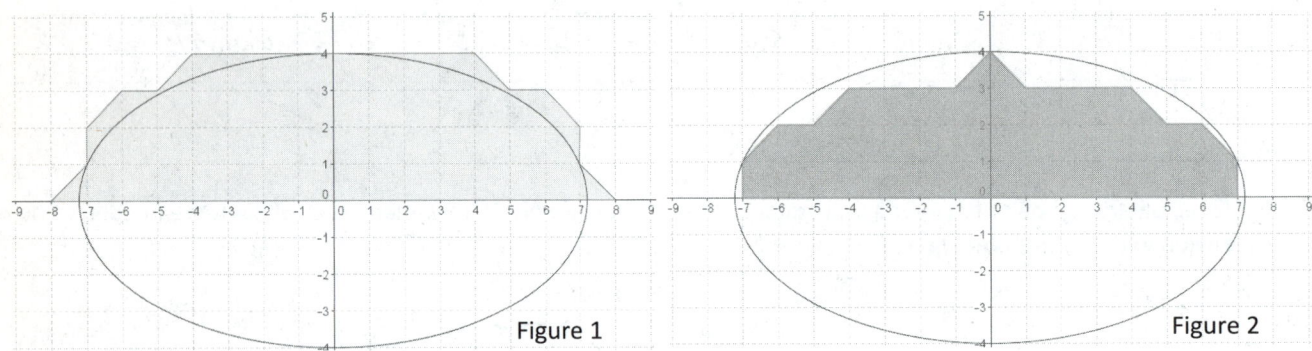

Figure 1   Figure 2

   a. Which polygonal region has a greater area? Explain your reasoning.

   b. Which, if either, of the polygonal regions do you believe is closer in area to the region inside the ellipse and above the x-axis?

3. Figures 1 and 2 below show two polygonal regions used to approximate the area of the region inside a parabola and above the x-axis.

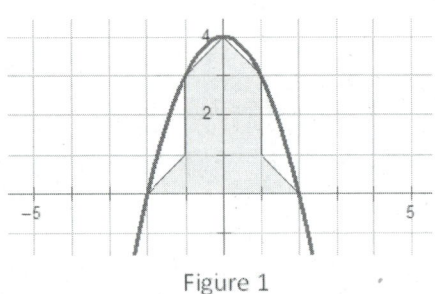 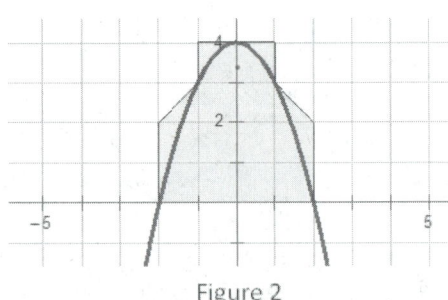

Figure 1   Figure 2

   a. Use the shaded polygonal region in Figure 1 to give a lower estimate of the area a under the curve and above the x-axis.

   b. Use the shaded polygonal region to give an upper estimate of the area a under the curve and above the x-axis.

   c. Use (a) and (b) to give an average estimate of the area a.

EUREKA
MATH™

4. Problem 4 is an extension of Problem 3. Using the diagram, draw grid lines to represent each $\frac{1}{2}$ unit.

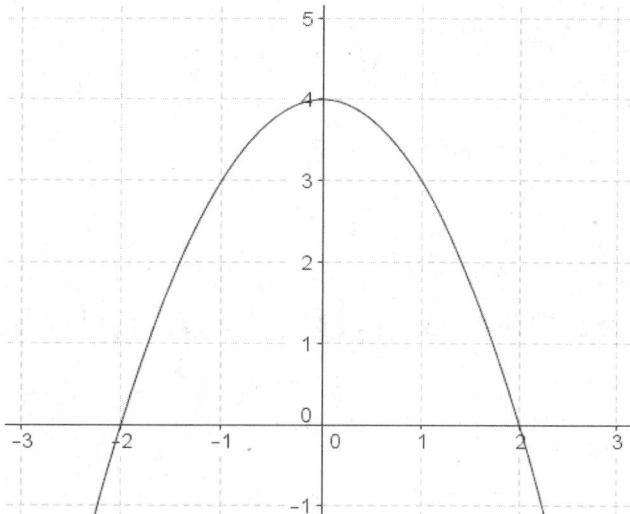

a. What do the new grid lines divide each unit square into?

b. Use the squares described in part (a) to determine a lower estimate of area a in the diagram.

c. Use the squares described in part (a) to determine an upper estimate of area a in the diagram.

d. Calculate an average estimate of the area under the curve and above the x-axis based on your upper and lower estimates in parts (b) and (c).

e. Do you think your average estimate in Problem 4 is more or less precise than your estimate from Problem 3? Explain.

This page intentionally left blank

Lesson 2: Properties of Area

Classwork

Exploratory Challenge/Exercises 1–4

1.   Two congruent triangles are shown below.

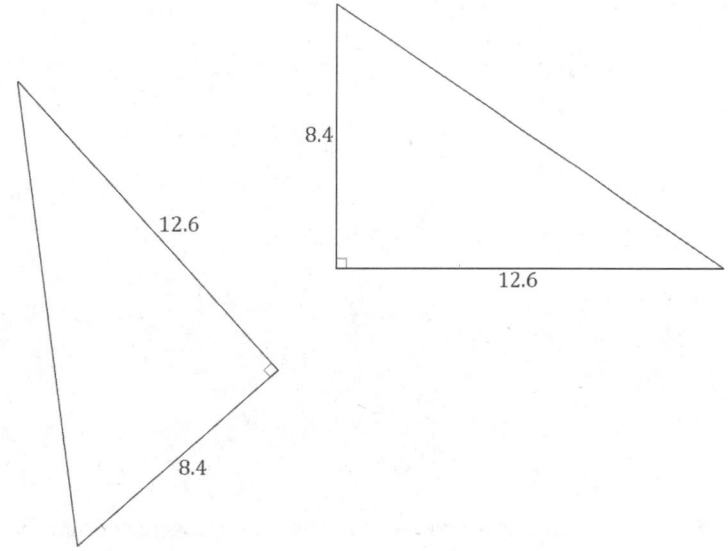

a.   Calculate the area of each triangle.

b.   Circle the transformations that, if applied to the first triangle, would always result in a new triangle with the same area:

Translation   Rotation   Dilation   Reflection

c.   Explain your answer to part (b).

2.

a.   Calculate the area of the shaded figure below.

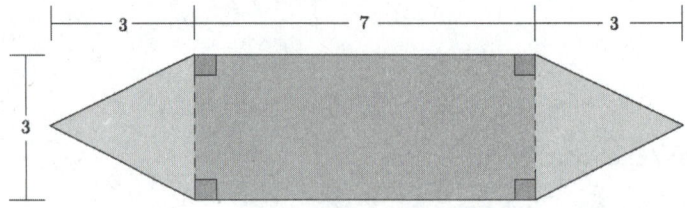

b.   Explain how you determined the area of the figure.

3.   Two triangles △ ABC and △ DEF are shown below. The two triangles overlap forming △ DGC.

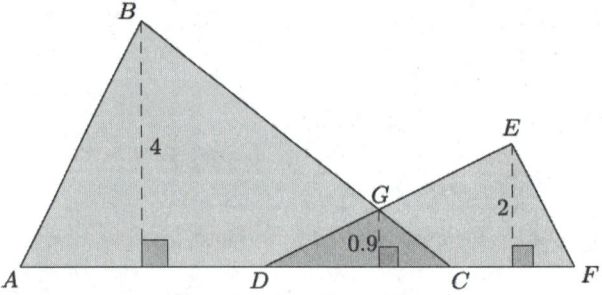

a.   The base of figure $ABGEF$ is composed of segments of the following lengths: $AD = 4$, $DC = 3$, and $CF = 2$.
     Calculate the area of the figure $ABGEF$.

EUREKA
MATH™

© 2015 Great Minds. eureka-math.org
GEO-M3-SE-B2-1.3.0-10.2015

b.   Explain how you determined the area of the figure.

4.   A rectangle with dimensions 21.6 × 12 has a right triangle with a base 9.6 and a height of 7.2 cut out of the rectangle.

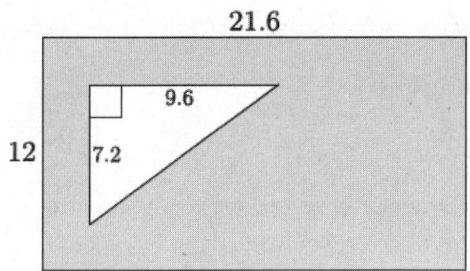

a.   Find the area of the shaded region.

b.   Explain how you determined the area of the shaded region.

Lesson Summary

SET (description): A *set* is a well-defined collection of objects called *elements* or *members* of the set.

SUBSET: A set A is a *subset* of a set B if every element of A is also an element of B. The notation $A \subseteq B$ indicates that the set A is a subset of set B.

UNION: The *union* of A and B is the set of all objects that are either elements of A or of B, or of both. The union is denoted $A \cup B$.

INTERSECTION: The *intersection* of A and B is the set of all objects that are elements of A and also elements of B. The intersection is denoted $A \cap B$.

Problem Set

1. Two squares with side length 5 meet at a vertex and together with segment AB form a triangle with base 6 as shown. Find the area of the shaded region.

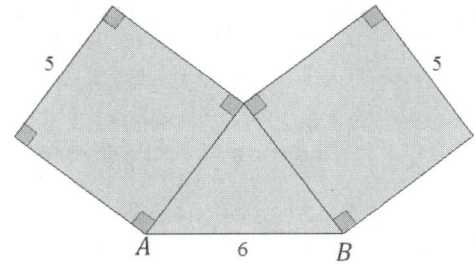

2. If two 2×2 square regions S_1 and S_2 meet at midpoints of sides as shown, find the area of the square region, $S_1 \cup S_2$.

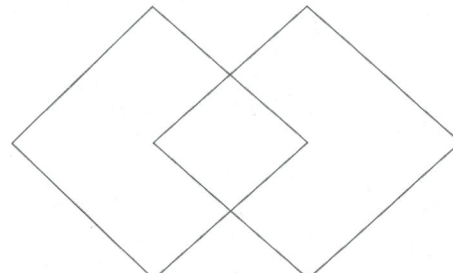

EUREKA
MATH™

3. The figure shown is composed of a semicircle and a non-overlapping equilateral triangle, and contains a hole that is also composed of a semicircle and a non-overlapping equilateral triangle. If the radius of the larger semicircle is 8, and the radius of the smaller semicircle is $\frac{1}{3}$ that of the larger semicircle, find the area of the figure.

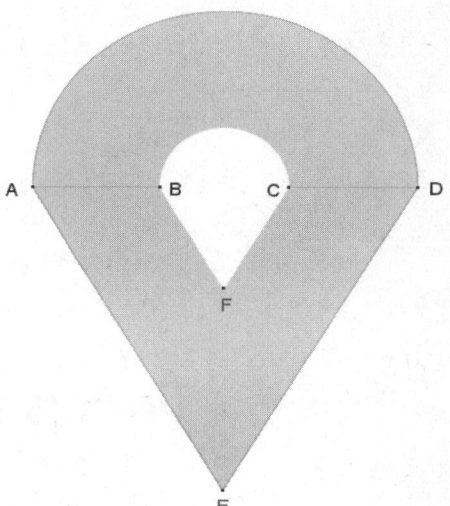

4. Two square regions A and B each have Area(8). One vertex of square B is the center point of square A. Can you find the area of $A \cup B$ and $A \cap B$ without any further information? What are the possible areas?

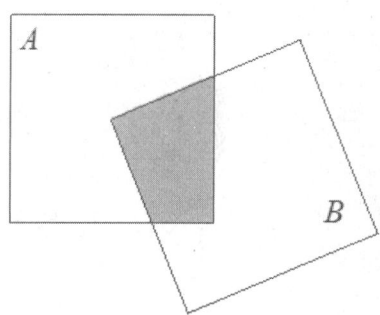

5. Four congruent right triangles with leg lengths a and b and hypotenuse length c are used to enclose the green region in Figure 1 with a square and then are rearranged inside the square leaving the green region in Figure 2.

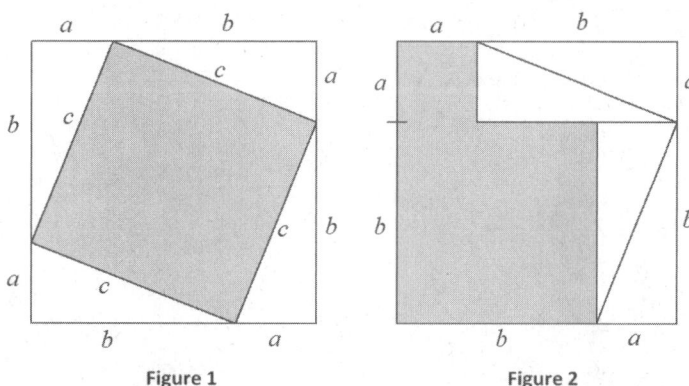

Figure 1   Figure 2

a. Use Property 4 to explain why the green region in Figure 1 has the same area as the green region in Figure 2.

b. Show that the green region in Figure 1 is a square, and compute its area.

c. Show that the green region in Figure 2 is the union of two non-overlapping squares, and compute its area.

d. How does this prove the Pythagorean theorem?

This page intentionally left blank

Lesson 3: The Scaling Principle for Area

Exploratory Challenge

Complete parts (i)–(iii) of the table for each of the figures in questions (a)–(d): (i) Determine the area of the figure (pre-image), (ii) determine the scaled dimensions of the figure based on the provided scale factor, and (iii) determine the area of the dilated figure. Then, answer the question that follows.

In the final column of the table, find the value of the ratio of the area of the similar figure to the area of the original figure.

	(i) Area of Original Figure	Scale Factor	(ii) Dimensions of Similar Figure	(iii) Area of Similar Figure	Ratio of Areas $\text{Area}_{\text{similar}} : \text{Area}_{\text{original}}$
a.		3			
b.		2			
c.		$\dfrac{1}{2}$			
d.		$\dfrac{3}{2}$			

a.

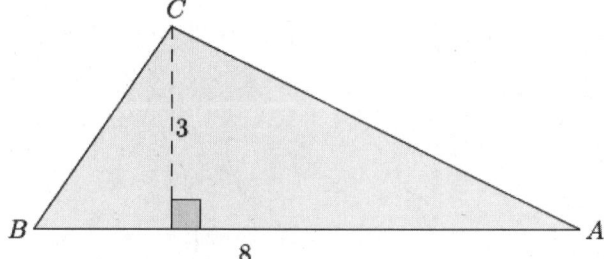

b.

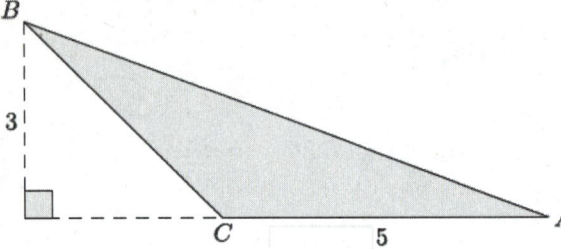

c.

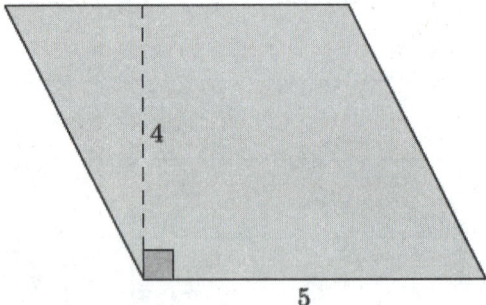

d.

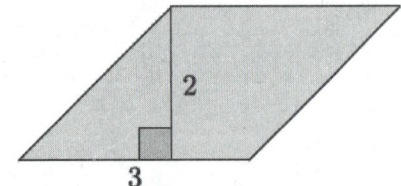

Lesson 3: The Scaling Principle for Area

EUREKA MATH

e. Make a conjecture about the relationship between the areas of the original figure and the similar figure with respect to the scale factor between the figures.

THE SCALING PRINCIPLE FOR TRIANGLES:

THE SCALING PRINCIPLE FOR POLYGONS:

Exercises 1–2

1. Rectangles A and B are similar and are drawn to scale. If the area of rectangle A is 88 mm^2, what is the area of rectangle B?

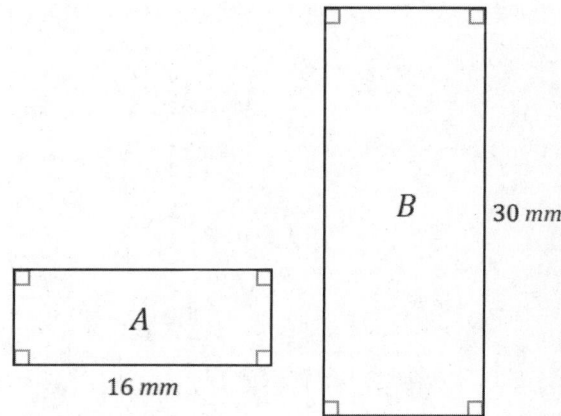

Lesson 3:   The Scaling Principle for Area

S.17

EUREKA
MATH™

© 2015 Great Minds. eureka-math.org
GEO-M3-SE-B2-1.3.0-10.2015

2. Figures E and F are similar and are drawn to scale. If the area of figure E is 120 mm², what is the area of figure F?

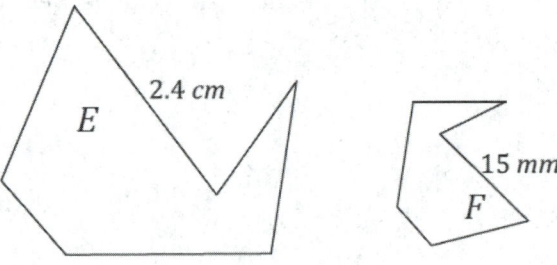

THE SCALING PRINCIPLE FOR AREA:

EUREKA
MATH™

Lesson Summary

THE SCALING PRINCIPLE FOR TRIANGLES: If similar triangles S and T are related by a scale factor of r, then the respective areas are related by a factor of r^2.

THE SCALING PRINCIPLE FOR POLYGONS: If similar polygons P and Q are related by a scale factor of r, then their respective areas are related by a factor of r^2.

THE SCALING PRINCIPLE FOR AREA: If similar figures A and B are related by a scale factor of r, then their respective areas are related by a factor of r^2.

Problem Set

1. A rectangle has an area of 18. Fill in the table below by answering the questions that follow. Part of the first row has been completed for you.

1	2	3	4	5	6
Original Dimensions	Original Area	Scaled Dimensions	Scaled Area	$\dfrac{\text{Scaled Area}}{\text{Original Area}}$	Area ratio in terms of the scale factor
18×1	18	$9 \times \dfrac{1}{2}$	$\dfrac{9}{2}$		

a. List five unique sets of dimensions of your choice of your choice for a rectangle with an area of 18, and enter them in column 1.

b. If the given rectangle is dilated from a vertex with a scale factor of $\dfrac{1}{2}$, what are the dimensions of the images of each of your rectangles? Enter the scaled dimensions in column 3.

c. What are the areas of the images of your rectangles? Enter the areas in column 4.

d. How do the areas of the images of your rectangles compare to the area of the original rectangle? Write the value of each ratio in simplest form in column 5.

e. Write the values of the ratios of area entered in column 5 in terms of the scale factor $\dfrac{1}{2}$. Enter these values in column 6.

f. If the areas of two unique rectangles are the same, x, and both figures are dilated by the same scale factor r, what can we conclude about the areas of the dilated images?

2. Find the ratio of the areas of each pair of similar figures. The lengths of corresponding line segments are shown.

    a.

    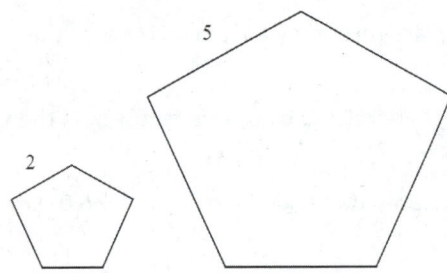

    b.

    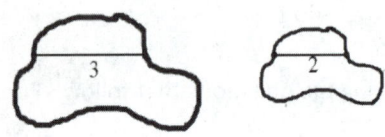

    c.

3. In △ ABC, line segment DE connects two sides of the triangle and is parallel to line segment BC. If the area of △ ABC is 54 and $BC = 3DE$, find the area of △ ADE.

    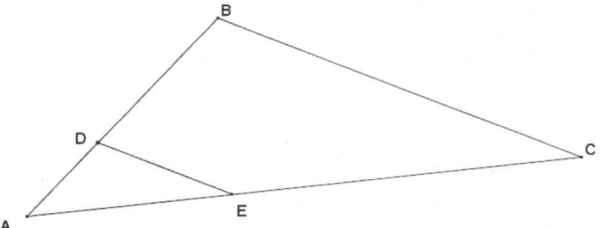

4. The small star has an area of 5. The large star is obtained from the small star by stretching by a factor of 2 in the horizontal direction and by a factor of 3 in the vertical direction. Find the area of the large star.

EUREKA
MATH™

5. A piece of carpet has an area of 50 yd². How many square inches will this be on a scale drawing that has 1 in. represent 1 yd.?

6. An isosceles trapezoid has base lengths of 12 in. and 18 in. If the area of the larger shaded triangle is 72 in², find the area of the smaller shaded triangle.

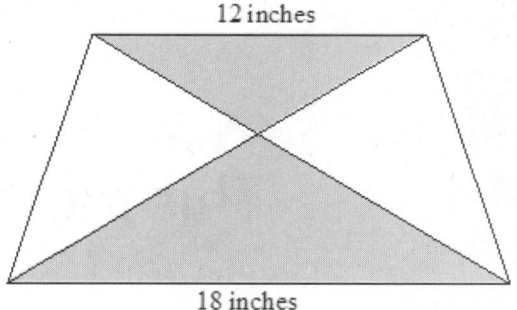

7. Triangle ABO has a line segment $A'B'$ connecting two of its sides so that $\overline{A'B'} \parallel \overline{AB}$. The lengths of certain segments are given. Find the ratio of the area of $\triangle OA'B'$ to the area of the quadrilateral $ABB'A'$.

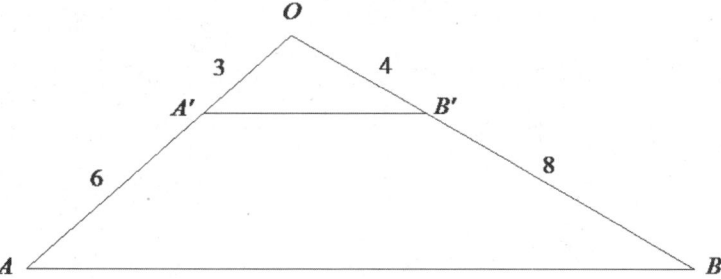

8. A square region S is scaled parallel to one side by a scale factor of r, $r \neq 0$, and is scaled in a perpendicular direction by a scale factor of one-third of r to yield its image S'. What is the ratio of the area of S to the area of S'?

9. Figure T' is the image of figure T that has been scaled horizontally by a scale factor of 4 and vertically by a scale factor of $\frac{1}{3}$. If the area of T' is 24 square units, what is the area of figure T?

10. What is the effect on the area of a rectangle if …
    a. Its height is doubled and base left unchanged?
    b. Its base and height are both doubled?
    c. Its base is doubled and height cut in half?

This page intentionally left blank

Lesson 4: Proving the Area of a Disk

Opening Exercise

The following image is of a regular hexagon inscribed in circle C with radius r. Find a formula for the area of the hexagon in terms of the length of a side, s, and the distance from the center to a side.

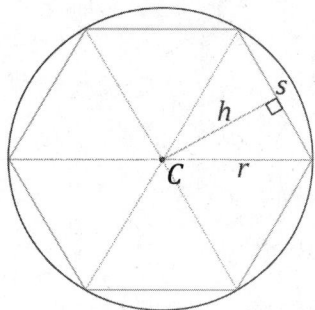

a.   Begin to approximate the area of a circle using inscribed polygons.

How well does a square approximate the area of a disk? Create a sketch of P_4 (a regular polygon with 4 sides, a square) in the following circle. Shade in the area of the disk that is not included in P_4.

b.   Next, create a sketch of P_8 in the following circle.

c.   Indicate which polygon has a greater area.

$$\text{Area}(P_4) \rule{2cm}{0.4pt} \text{Area}(P_8)$$

d.   Will the area of inscribed regular polygon P_{16} be greater or less than the area of P_8? Which is a better approximation of the area of the disk?

e.   We noticed that the area of P_4 was less than the area of P_8 and that the area of P_8 was less than the area of P_{16}. In other words, $\text{Area}(P_n)$ _____ $\text{Area}(P_{2n})$. Why is this true?

EUREKA
MATH™

f.   Now we will approximate the area of a disk using circumscribed (outer) polygons.

Now circumscribe, or draw a square on the outside of, the following circle such that each side of the square intersects the circle at one point. We will denote each of our outer polygons with prime notation; we are sketching P'_4 here.

g.   Create a sketch of P'_8.

h.   Indicate which polygon has a greater area.

$$\text{Area}(P'_4) \ \underline{\hspace{2cm}} \ \text{Area}(P'_8)$$

i.   Which is a better approximation of the area of the circle, P'_4 or P'_8? Explain why.

j.   How will Area(P'_n) compare to Area(P'_{2n})? Explain.

LIMIT (description): Given an infinite sequence of numbers, $a_1, a_2, a_3, \ldots$, to say that *the limit of the sequence is A* means, roughly speaking, that when the index n is very large, then a_n is very close to A. This is often denoted as, "As $n \to \infty$, $a_n \to A$."

AREA OF A CIRCLE (description): The *area of a circle* is the limit of the areas of the inscribed regular polygons as the number of sides of the polygons approaches infinity.

Problem Set

1. Describe a method for obtaining closer approximations of the area of a circle. Draw a diagram to aid in your explanation.

2. What is the radius of a circle whose circumference is π?

3. The side of a square is 20 cm long. What is the circumference of the circle when …
   a. The circle is inscribed within the square?
   b. The square is inscribed within the circle?

4. The circumference of circle C_1 is 9 cm, and the circumference of C_2 is 2π cm. What is the value of the ratio of the areas of C_1 to C_2?

5. The circumference of a circle and the perimeter of a square are each 50 cm. Which figure has the greater area?

6. Let us define π to be the circumference of a circle whose diameter is 1.

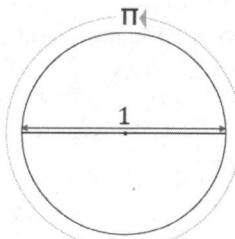

We are going to show why the circumference of a circle has the formula $2\pi r$. Circle C_1 below has a diameter of $d = 1$, and circle C_2 has a diameter of $d = 2r$.

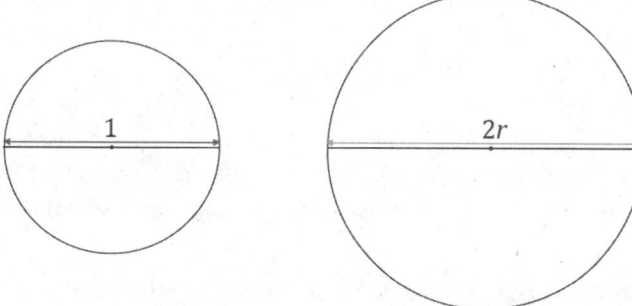

   a. All circles are similar (proved in Module 2). What scale factor of the similarity transformation takes C_1 to C_2?
   b. Since the circumference of a circle is a one-dimensional measurement, the value of the ratio of two circumferences is equal to the value of the ratio of their respective diameters. Rewrite the following equation by filling in the appropriate values for the diameters of C_1 and C_2:

   $$\frac{\text{Circumference}(C_2)}{\text{Circumference}(C_1)} = \frac{\text{diameter}(C_2)}{\text{diameter}(C_1)}.$$

c.   Since we have defined π to be the circumference of a circle whose diameter is 1, rewrite the above equation using this definition for C_1.

d.   Rewrite the equation to show a formula for the circumference of C_2.

e.   What can we conclude?

7.

a.   Approximate the area of a disk of radius 1 using an inscribed regular hexagon. What is the percent error of the approximation?

(Remember that percent error is the absolute error as a percent of the exact measurement.)

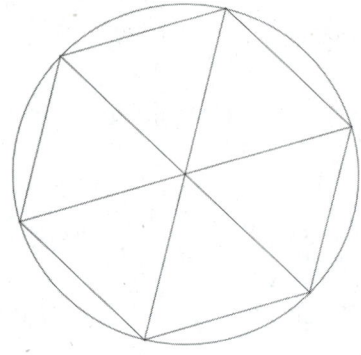

b.   Approximate the area of a circle of radius 1 using a circumscribed regular hexagon. What is the percent error of the approximation?

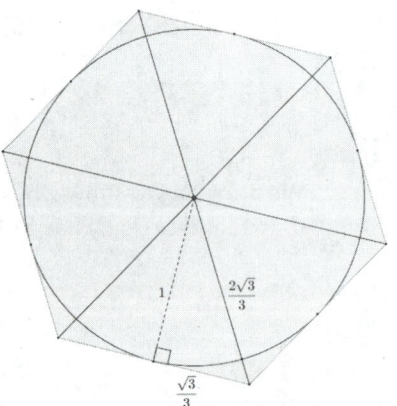

c.   Find the average of the approximations for the area of a circle of radius 1 using inscribed and circumscribed regular hexagons. What is the percent error of the average approximation?

8.   A regular polygon with n sides each of length s is inscribed in a circle of radius r. The distance h from the center of the circle to one of the sides of the polygon is over 98% of the radius. If the area of the polygonal region is 10, what can you say about the area of the circumscribed regular polygon with n sides?

Lesson 5: Three-Dimensional Space

Exercise

The following three-dimensional right rectangular prism has dimensions $3 \times 4 \times 5$. Determine the length of $\overline{AC'}$. Show a full solution.

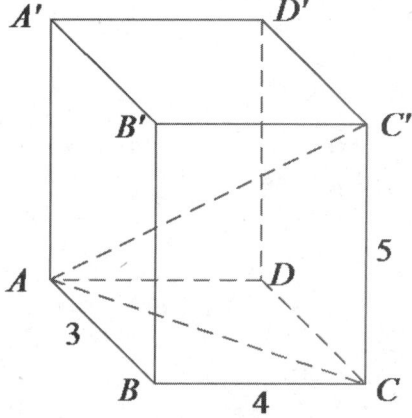

Exploratory Challenge

Table 1: Properties of Points, Lines, and Planes in Three-Dimensional Space

	Property	**Diagram**		
1	Two points P and Q determine a distance PQ, a line segment PQ, a ray PQ, a vector PQ, and a line PQ.			
2	Three non-collinear points A, B, and C determine a plane ABC and, in that plane, determine a triangle ABC.	Given a picture of the plane below, sketch a triangle in that plane.		
3	Two lines either meet in a single point, or they do not meet. Lines that do not meet and lie in a plane are called *parallel*. *Skew* lines are lines that do not meet and are not parallel.	(a) Sketch two lines that meet in a single point.	(b) Sketch lines that do not meet and lie in the same plane; i.e., sketch parallel lines.	(c) Sketch a pair of skew lines.

EUREKA
MATH™

4	Given a line ℓ and a point not on ℓ, there is a unique line through the point that is parallel to ℓ.			
5	Given a line ℓ and a plane P, then ℓ lies in P, ℓ meets P in a single point, or ℓ does not meet P, in which case we say ℓ is *parallel* to P. (Note: This implies that if two points lie in a plane, then the line determined by the two points is also in the plane.)	(a) Sketch a line ℓ that lies in plane P.	(b) Sketch a line ℓ that meets P in a single point.	(c) Sketch a line ℓ that does not meet P; i.e., sketch a line ℓ parallel to P.
6	Two planes either meet in a line, or they do not meet, in which case we say the planes are *parallel*.	(a) Sketch two planes that meet in a line.	(b) Sketch two planes that are parallel.	

7	Two rays with the same vertex form an angle. The angle lies in a plane and can be measured by degrees.	Sketch the example in the following plane:
8	Two lines are *perpendicular* if they meet, and any of the angles formed between the lines is a *right angle*. Two segments or rays are perpendicular if the lines containing them are perpendicular lines.	
9	A line ℓ is perpendicular to a plane P if they meet in a single point, and the plane contains two lines that are perpendicular to ℓ, in which case every line in P that meets ℓ is perpendicular to ℓ. A segment or ray is perpendicular to a plane if the line determined by the ray or segment is perpendicular to the plane.	Draw an example of a line that is perpendicular to a plane. Draw several lines that lie in the plane that pass through the point where the perpendicular line intersects the plane.
10	Two planes perpendicular to the same line are parallel.	

Lesson 5: Three-Dimensional Space

EUREKA MATH™

11	Two lines perpendicular to the same plane are parallel.	Sketch an example that illustrates this statement using the following plane:
12	Any two line segments connecting parallel planes have the same length if they are each perpendicular to one (and hence both) of the planes.	Sketch an example that illustrates this statement using parallel planes P and Q.
13	The *distance between a point and a plane* is the length of the perpendicular segment from the point to the plane. The distance is defined to be zero if the point is on the plane. The *distance between two planes* is the distance from a point in one plane to the other.	Sketch the segment from A that can be used to measure the distance between A and the plane P. •A

> ## Lesson Summary
>
> **SEGMENT:** The *segment between points A and B* is the set consisting of A, B, and all points on $\overleftrightarrow{AB}$ between A and B. The segment is denoted by $\overline{AB}$, and the points A and B are called the *endpoints.*
>
> **LINE PERPENDICULAR TO A PLANE:** A line L intersecting a plane E at a point P is said to be *perpendicular to the plane E* if L is perpendicular to every line that (1) lies in E and (2) passes through the point P. A segment is said to be perpendicular to a plane if the line that contains the segment is perpendicular to the plane.

Problem Set

1. Indicate whether each statement is always true (A), sometimes true (S), or never true (N).
   a. If two lines are perpendicular to the same plane, the lines are parallel.
   b. Two planes can intersect in a point.
   c. Two lines parallel to the same plane are perpendicular to each other.
   d. If a line meets a plane in one point, then it must pass through the plane.
   e. Skew lines can lie in the same plane.
   f. If two lines are parallel to the same plane, the lines are parallel.
   g. If two planes are parallel to the same line, they are parallel to each other.
   h. If two lines do not intersect, they are parallel.

2. Consider the right hexagonal prism whose bases are regular hexagonal regions. The top and the bottom hexagonal regions are called the *base faces*, and the side rectangular regions are called the *lateral faces*.
   a. List a plane that is parallel to plane $C'D'E'$.
   b. List all planes shown that are not parallel to plane CDD'.
   c. Name a line perpendicular to plane ABC.
   d. Explain why $AA' = CC'$.
   e. Is $\overleftrightarrow{AB}$ parallel to $\overleftrightarrow{DE}$? Explain.
   f. Is $\overleftrightarrow{AB}$ parallel to $\overleftrightarrow{C'D'}$? Explain.
   g. Is $\overleftrightarrow{AB}$ parallel to $\overleftrightarrow{D'E'}$? Explain.
   h. If $\overline{BC'}$ and $\overline{C'F'}$ are perpendicular, then is $\overleftrightarrow{BC}$ perpendicular to plane $C'A'F'$? Explain.
   i. One of the following statements is false. Identify which statement is false, and explain why.
      (i) $\overleftrightarrow{BB'}$ is perpendicular to $\overleftrightarrow{B'C'}$.
      (ii) $\overleftrightarrow{EE'}$ is perpendicular to $\overleftrightarrow{EF}$.
      (iii) $\overleftrightarrow{CC'}$ is perpendicular to $\overleftrightarrow{E'F'}$.
      (iv) $\overleftrightarrow{BC}$ is parallel to $\overleftrightarrow{F'E'}$.

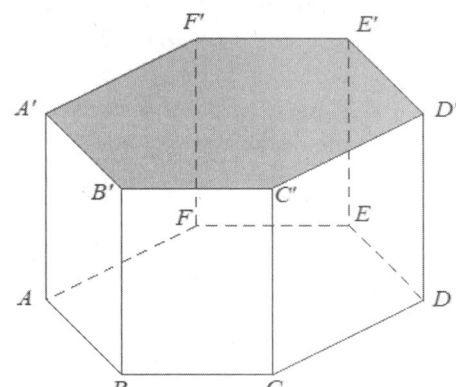

3. In the following figure, △ ABC is in plane P, △ DEF is in plane Q, and BCFE is a rectangle. Which of the following statements are true?

    a. $\overline{BE}$ is perpendicular to plane Q.

    b. $BF = CE$

    c. Plane P is parallel to plane Q.

    d. $\triangle ABC \cong \triangle DEF$

    e. $AE = AF$

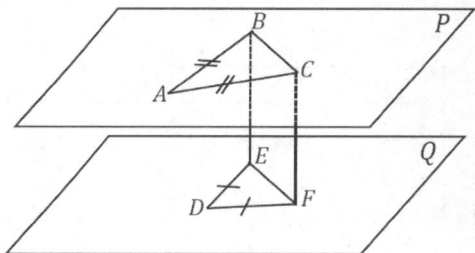

4. Challenge: The following three-dimensional right rectangular prism has dimensions $a \times b \times c$. Determine the length of $\overline{AC'}$.

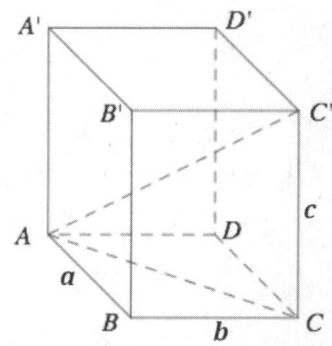

5. A line ℓ is perpendicular to plane P. The line and plane meet at point C. If A is a point on ℓ different from C, and B is a point on P different from C, show that $AC < AB$.

6. Given two distinct parallel planes P and R, $\overleftrightarrow{EF}$ in P with $EF = 5$, point G in R, $m\angle GEF = 90°$, and $m\angle EFG = 60°$, find the minimum and maximum distances between planes P and R, and explain why the actual distance is unknown.

7. The diagram below shows a right rectangular prism determined by vertices A, B, C, D, E, F, G, and H. Square ABCD has sides with length 5, and $AE = 9$. Find DF.

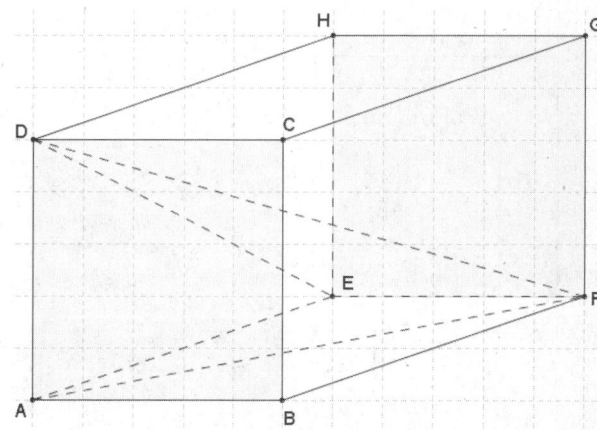

Table 2: Properties of Points, Lines, and Planes in Three-Dimensional Space

	Property	Diagram
1		
2		
3		(a) (b) (c)
4		

5		(a)	(b)	(c)

6		(a)	(b)

7	

8	

9	
	ℓ *P*
10	
	ℓ *P* *Q*
11	
	ℓ *m*
12	
	B *D* *P* *A* *C* *Q* *AB = CD*
13	
	A *P*

EUREKA
MATH™

© 2015 Great Minds. eureka-math.org
GEO-M3-SE-B2-1.3.0-10.2015

Lesson 6: General Prisms and Cylinders and Their Cross-Sections

Classwork

Opening Exercise

Sketch a right rectangular prism.

RIGHT RECTANGULAR PRISM: Let E and E' be two parallel planes. Let B be a rectangular region[1] in the plane E. At each point P of B, consider $\overline{PP'}$ perpendicular to E, joining P to a point P' of the plane E'. The union of all these segments is called a *right rectangular prism*.

GENERAL CYLINDER: (See Figure 1.) Let E and E' be two parallel planes, let B be a region[2] in the plane E, and let L be a line that intersects E and E' but not B. At each point P of B, consider $\overline{PP'}$ parallel to L, joining P to a point P' of the plane E'. The union of all these segments is called a *general cylinder with base B*.

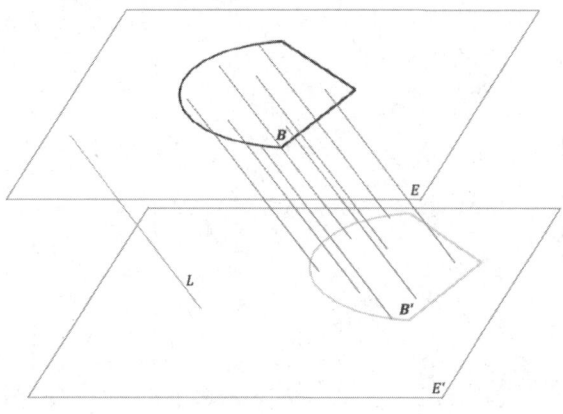

Figure 1

---

[1](Fill in the blank.) A rectangular region is the union of a rectangle and _____ .
[2]In Grade 8, a *region* refers to a *polygonal region* (triangle, quadrilateral, pentagon, and hexagon), a *circular region,* or regions that can be decomposed into such regions.

Discussion

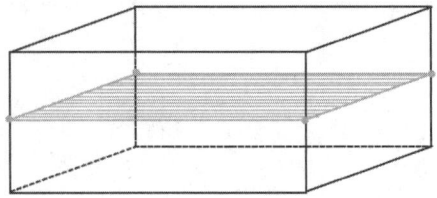 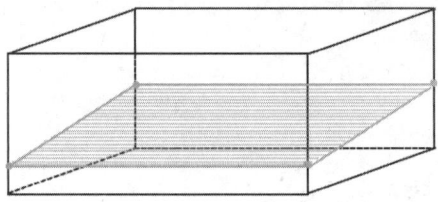

Figure 2

Example of a cross-section of a prism, where the intersection of a plane with the solid is parallel to the base.

Figure 3

A general intersection of a plane with a prism, which is sometimes referred to as a slice.

Exercise

Sketch the cross-section for the following figures:

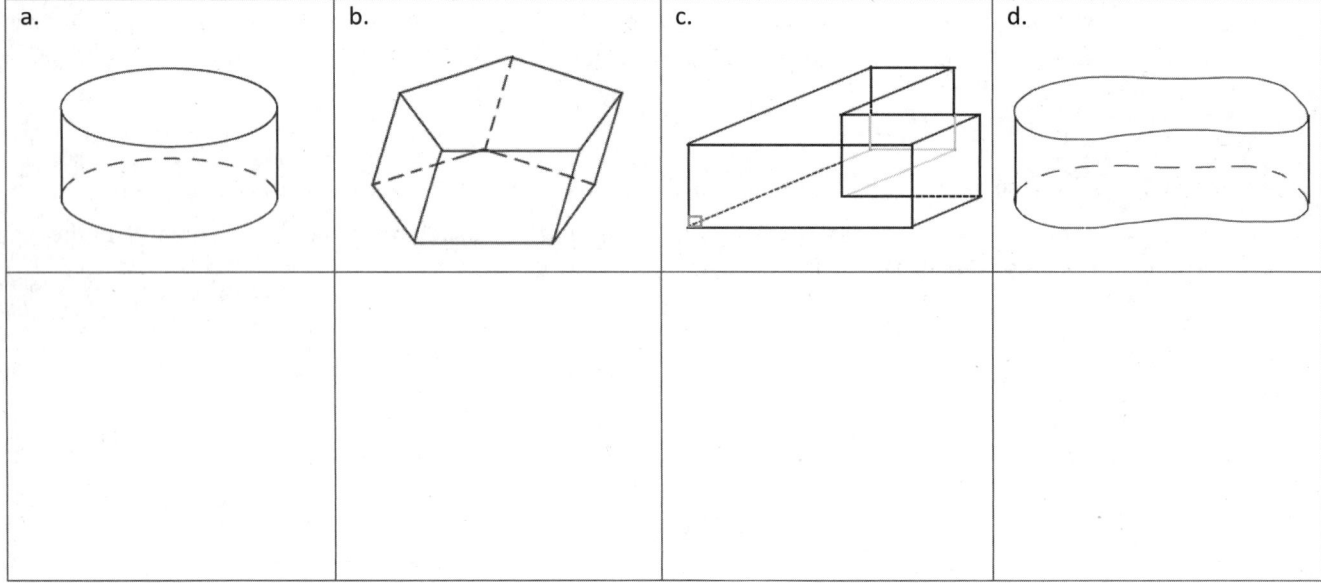

a.	b.	c.	d.

Lesson 6: General Prisms and Cylinders and Their Cross-Sections

EUREKA MATH™

Extension

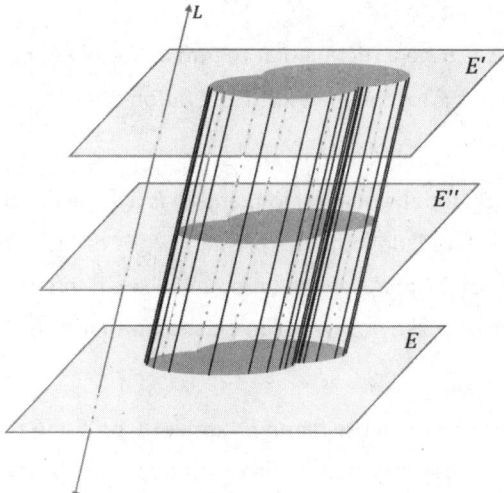

Figure 4

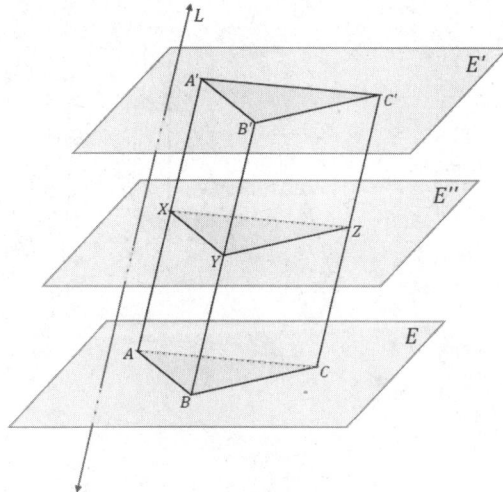

Figure 5

Lesson Summary

RIGHT RECTANGULAR PRISM: Let E and E' be two parallel planes. Let B be a rectangular region in the plane E. At each point P of B, consider $\overline{PP'}$ perpendicular to E, joining P to a point P' of the plane E'. The union of all these segments is called a *right rectangular prism*.

LATERAL EDGE AND FACE OF A PRISM: Suppose the base B of a prism is a polygonal region, and P_i is a vertex of B. Let P'_i be the corresponding point in B' such that $\overline{P_i P'_i}$ is parallel to the line L defining the prism. $\overline{P_i P'_i}$ is called a *lateral edge of the prism*. If $\overline{P_i P_{i+1}}$ is a base edge of the base B (a side of B), and F is the union of all segments PP' parallel to L for which P is in $\overline{P_i P_{i+1}}$ and P' is in B', then F is a *lateral face of the prism*. It can be shown that a lateral face of a prism is always a region enclosed by a parallelogram.

GENERAL CYLINDER: Let E and E' be two parallel planes, let B be a region in the plane E, and let L be a line that intersects E and E' but not B. At each point P of B, consider $\overline{PP'}$ parallel to L, joining P to a point P' of the plane E'. The union of all these segments is called a *general cylinder with base B*.

Problem Set

1. Complete each statement below by filling in the missing term(s).

   a. The following prism is called a(n) _____ prism.

   b. If $\overline{AA'}$ were perpendicular to the plane of the base, then the prism would be called a(n) _____ prism.

   c. The regions $ABCD$ and $A'B'C'D'$ are called the _____ of the prism.

   d. $\overline{AA'}$ is called a(n) _____.

   e. Parallelogram region $BB'C'C$ is one of four _____ _____.

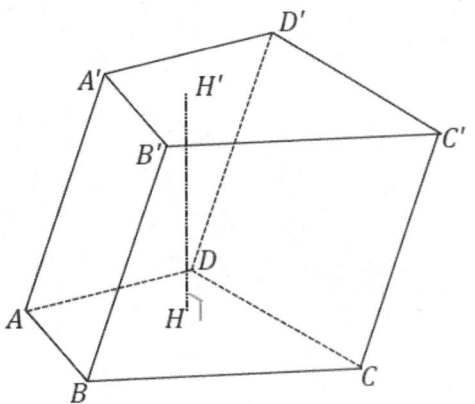

EUREKA
MATH™

2. The following right prism has trapezoidal base regions; it is a right trapezoidal prism. The lengths of the parallel edges of the base are 5 and 8, and the nonparallel edges are 4 and 6; the height of the trapezoid is 3.7. The lateral edge length DH is 10. Find the surface area of the prism.

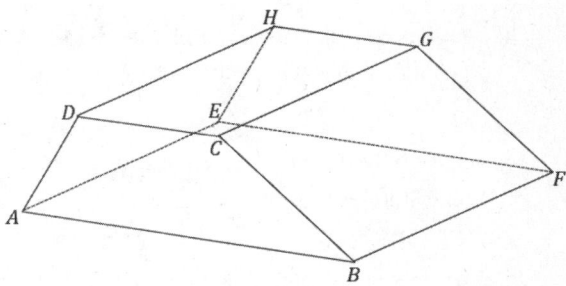

3. The base of the following right cylinder has a circumference of 5π and a lateral edge of 8. What is the radius of the base? What is the lateral area of the right cylinder?

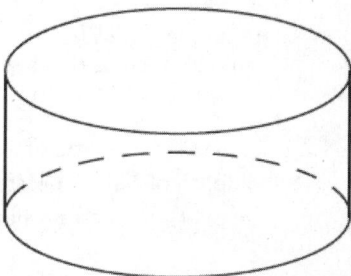

4. The following right general cylinder has a lateral edge of length 8, and the perimeter of its base is 27. What is the lateral area of the right general cylinder?

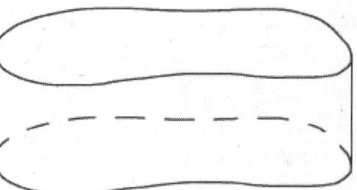

5. A right prism has base area 5 and volume 30. Find the prism's height, h.

6. Sketch the figures formed if the rectangular regions are rotated around the provided axis.

a.

b.

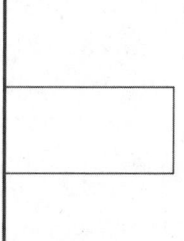

7. A cross-section is taken parallel to the bases of a general cylinder and has an area of 18. If the height of the cylinder is h, what is the volume of the cylinder? Explain your reasoning.

8. A general cylinder has a volume of 144. What is one possible set of dimensions of the base and height of the cylinder if all cross-sections parallel to its bases are...

    a. Rectangles?

    b. Right triangles?

    c. Circles?

9. A general hexagonal prism is given. If P is a plane that is parallel to the planes containing the base faces of the prism, how does P meet the prism?

10. Two right prisms have similar bases. The first prism has height 5 and volume 100. A side on the base of the first prism has length 2, and the corresponding side on the base of the second prism has length 3. If the height of the second prism is 6, what is its volume?

11. A tank is the shape of a right rectangular prism with base 2 ft. × 2 ft. and height 8 ft. The tank is filled with water to a depth of 6 ft. A person of height 6 ft. jumps in and stands on the bottom. About how many inches will the water be over the person's head? Make reasonable assumptions.

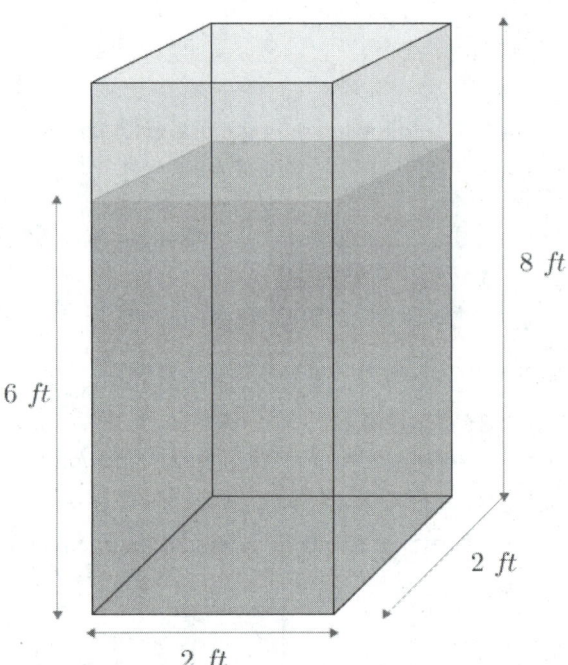

Lesson 6:   General Prisms and Cylinders and Their Cross-Sections   **EUREKA MATH**

Exploratory Challenge

Option 1

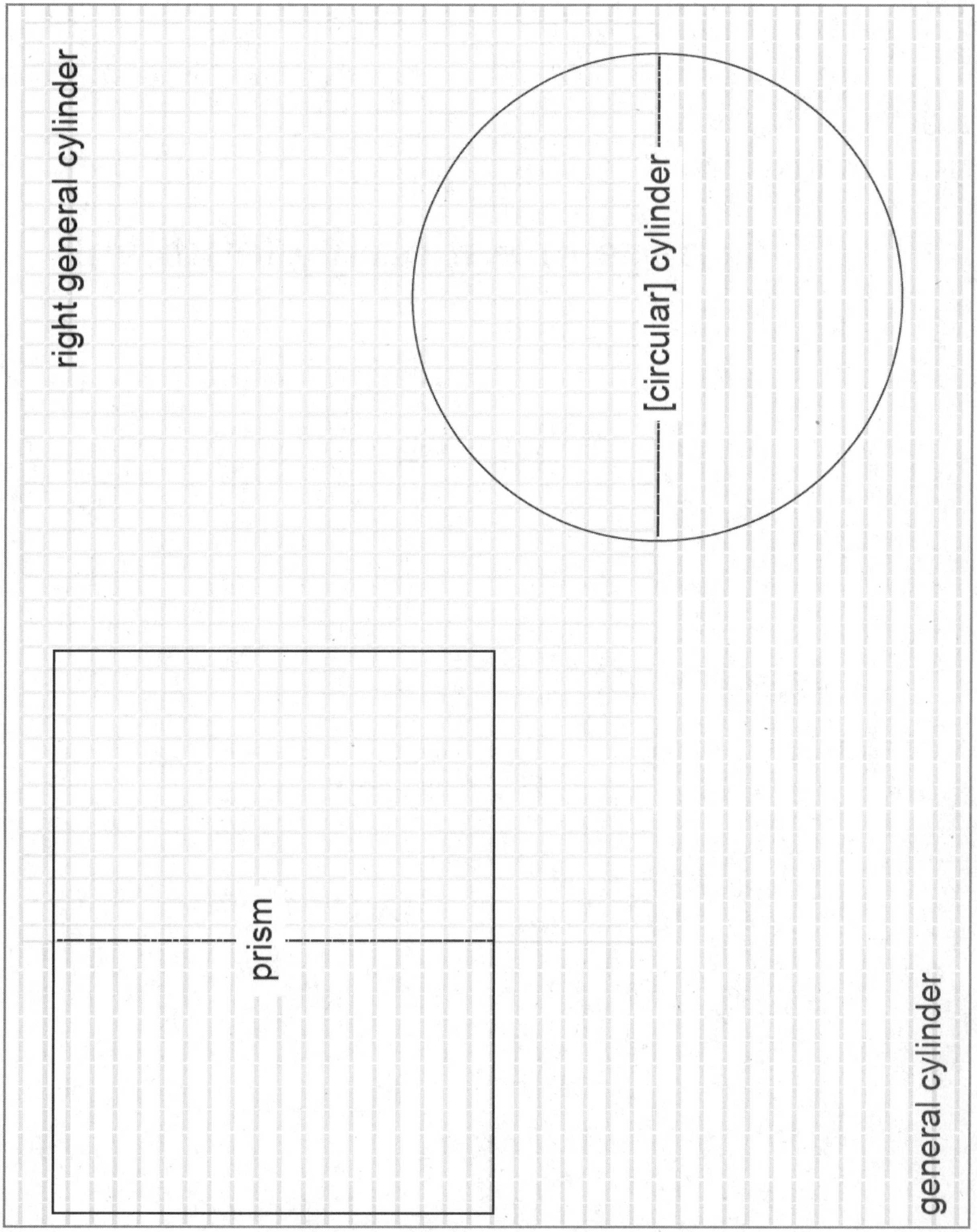

right general cylinder

[circular] cylinder

prism

general cylinder

This page intentionally left blank

Option 2

Figure and Description	Sketch of Figure	Sketch of Cross-Section
1. **General Cylinder** Let E and E' be two parallel planes, let B be a region in the plane E, and let L be a line that intersects E and E' but not B. At each point P of B, consider the segment $\overline{PP'}$ parallel to L, joining P to a point P' of the plane E'. The union of all these segments is called a *general cylinder with base B*.		
2. **Right General Cylinder** A general cylinder whose lateral edges are perpendicular to the bases.		
3. **Right Prism** A general cylinder whose lateral edges are perpendicular to a polygonal base.		
4. **Oblique Prism** A general cylinder whose lateral edges are not perpendicular to a polygonal base.		
5. **Right Cylinder** A general cylinder whose lateral edges are perpendicular to a circular base.		
6. **Oblique Cylinder** A general cylinder whose lateral edges are not perpendicular to a circular base.		

Option 3

	Figure and Description	Sketch of Figure	Sketch of Cross-Section
1.	**General Cylinder**	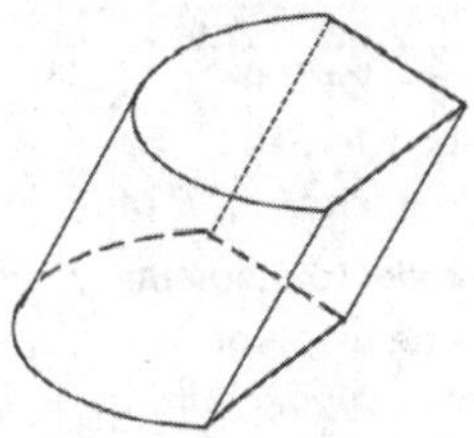	
2.	**Right General Cylinder**	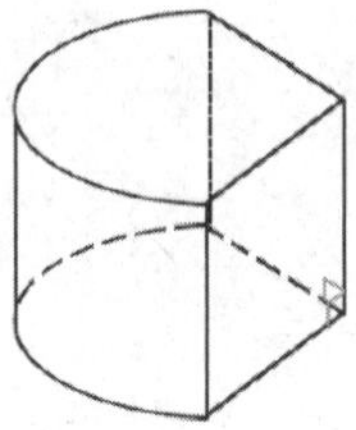	
3.	**Right Prism**	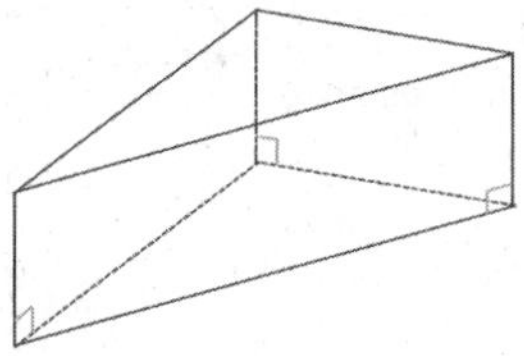	
4.	**Oblique Prism**	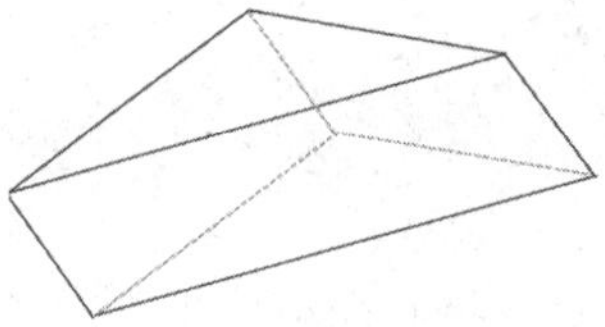	
5.	**Right Cylinder**	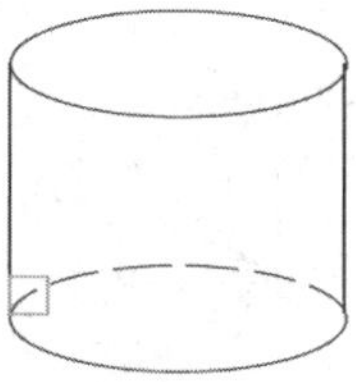	
6.	**Oblique Cylinder**	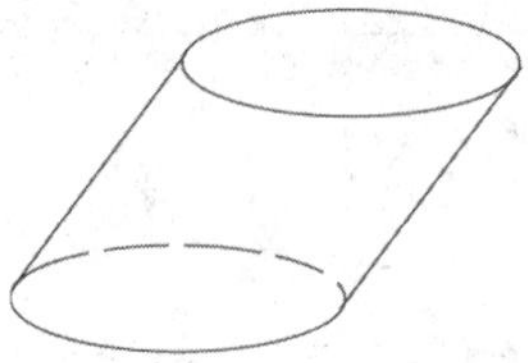	

EUREKA
MATH™

Lesson 7: General Pyramids and Cones and Their Cross-Sections

Classwork

Opening Exercise

Group the following images by shared properties. What defines each of the groups you have made?

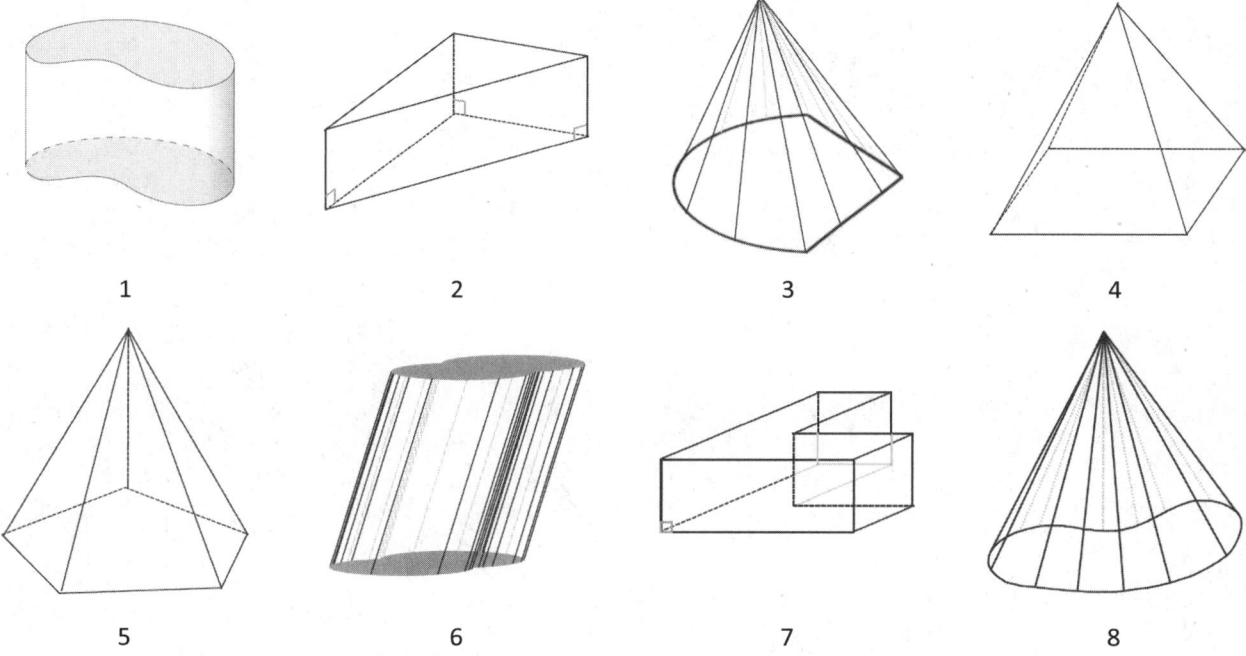

1   2   3   4

5   6   7   8

RECTANGULAR PYRAMID: Given a rectangular region B in a plane E and a point V not in E, the *rectangular pyramid with base B and vertex V* is the collection of all segments VP for any point P in B.

GENERAL CONE: Let B be a region in a plane E and V be a point not in E. The *cone with base B and vertex V* is the union of all segments VP for all points P in B (See Figures 1 and 2).

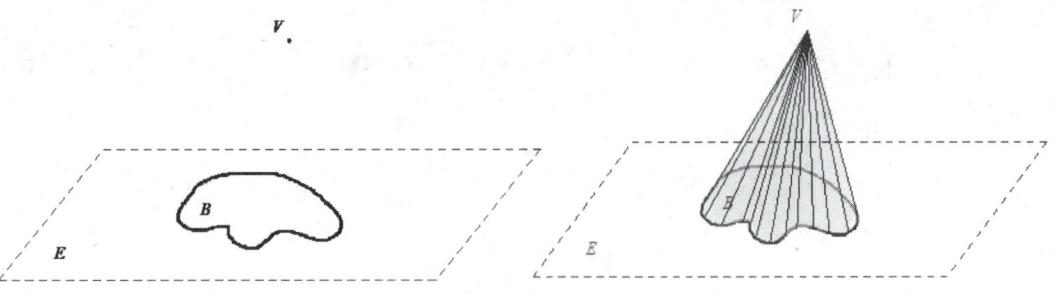

Figure 1

Figure 2

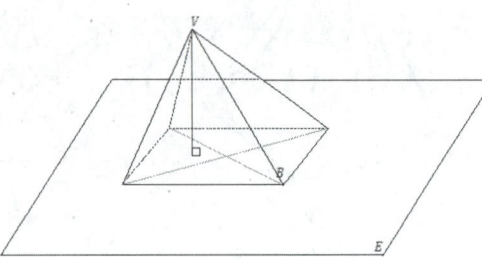

Figure 3

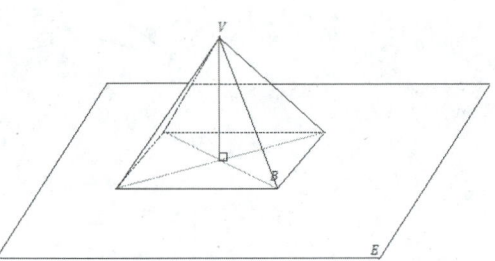

Figure 4

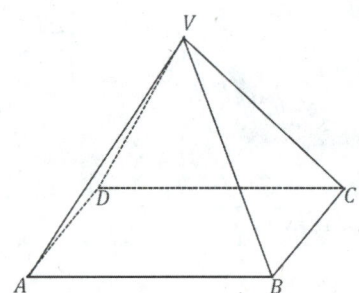

Figure 5

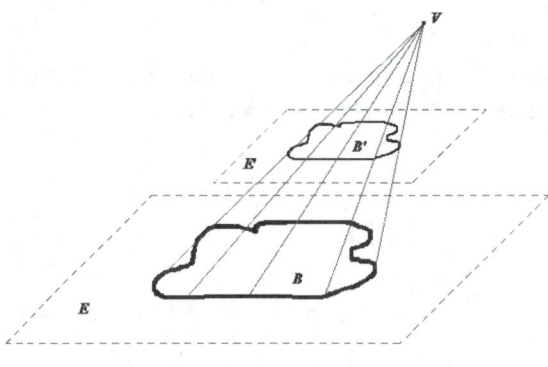

Figure 6

EUREKA
MATH™

Example 1

In the following triangular pyramid, a plane passes through the pyramid so that it is parallel to the base and results in the cross-section △ $A'B'C'$. If the area of △ ABC is 25 mm², what is the area of △ $A'B'C'$?

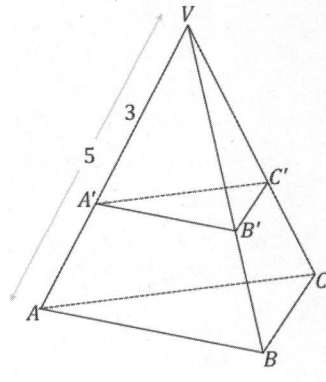

Example 2

In the following triangular pyramid, a plane passes through the pyramid so that it is parallel to the base and results in the cross-section △ $A'B'C'$. The altitude from V is drawn; the intersection of the altitude with the base is X, and the intersection of the altitude with the cross-section is X'. If the distance from X to V is 18 mm, the distance from X' to V is 12 mm, and the area of △ $A'B'C'$ is 28 mm², what is the area of △ ABC?

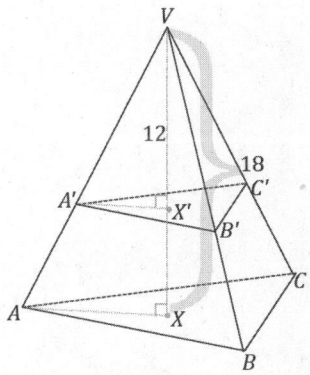

EUREKA
MATH™

Extension

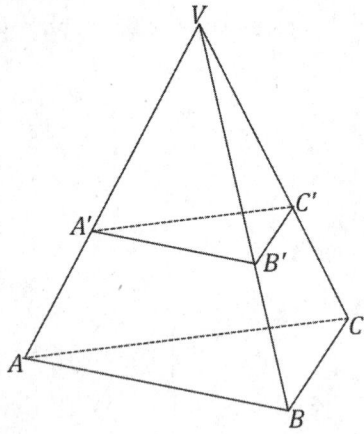

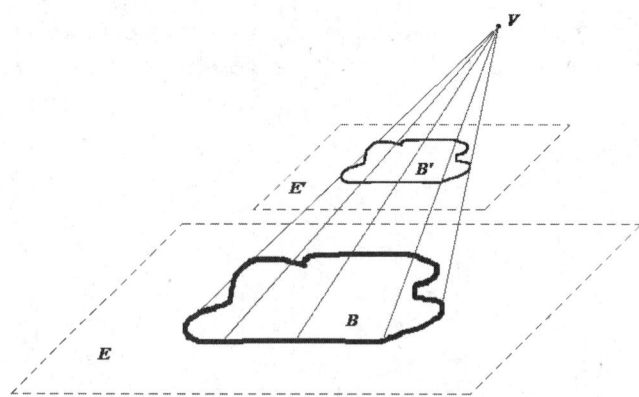

Exercise 1

The area of the base of a cone is 16, and the height is 10. Find the area of a cross-section that is distance 5 from the vertex.

Lesson 7:   General Pyramids and Cones and Their Cross-Sections

Example 3

GENERAL CONE CROSS-SECTION THEOREM: If two general cones have the same base area and the same height, then cross-sections for the general cones the same distance from the vertex have the same area.

State the theorem in your own words.

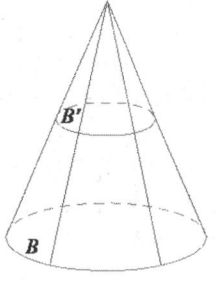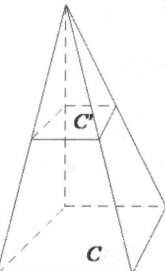

Figure 8

Use the space below to prove the *general cone cross-section theorem*.

Exercise 2

The following pyramids have equal altitudes, and both bases are equal in area and are coplanar. Both pyramids' cross-sections are also coplanar. If $BC = 3\sqrt{2}$ and $B'C' = 2\sqrt{3}$, and the area of $TUVWXYZ$ is 30 units^2, what is the area of cross-section $A'B'C'D'$?

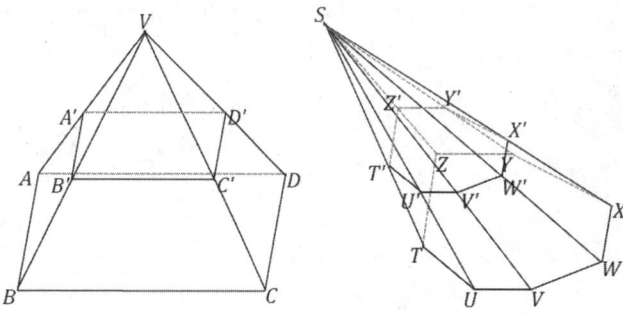

Problem Set

1. The base of a pyramid has area 4. A cross-section that lies in a parallel plane that is distance of 2 from the base plane has an area of 1. Find the height, h, of the pyramid.

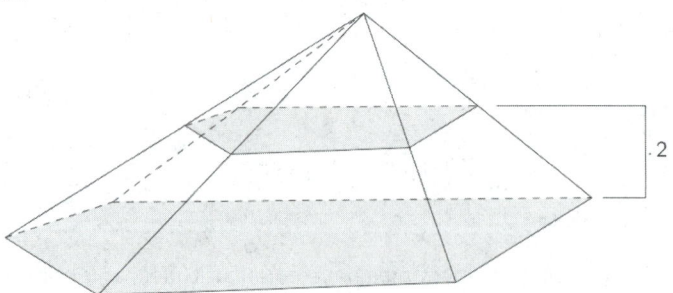

2. The base of a pyramid is a trapezoid. The trapezoidal bases have lengths of 3 and 5, and the trapezoid's height is 4. Find the area of the parallel slice that is three-fourths of the way from the vertex to the base.

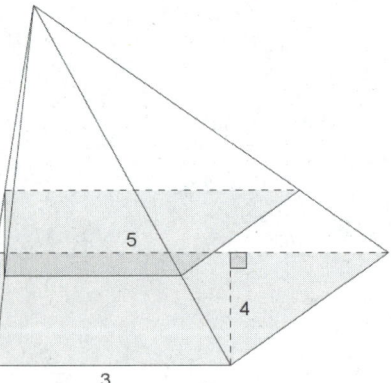

Lesson 7:   General Pyramids and Cones and Their Cross-Sections

EUREKA MATH™

3. A cone has base area 36 cm^2. A parallel slice 5 cm from the vertex has area 25 cm^2. Find the height of the cone.

4. Sketch the figures formed if the triangular regions are rotated around the provided axis:

    a. 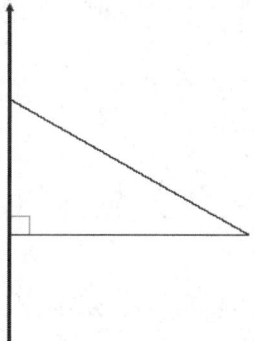   b.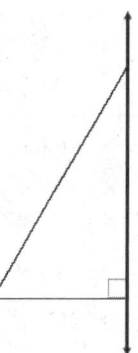

5. Liza drew the top view of a rectangular pyramid with two cross-sections as shown in the diagram and said that her diagram represents one, and only one, rectangular pyramid. Do you agree or disagree with Liza? Explain.

    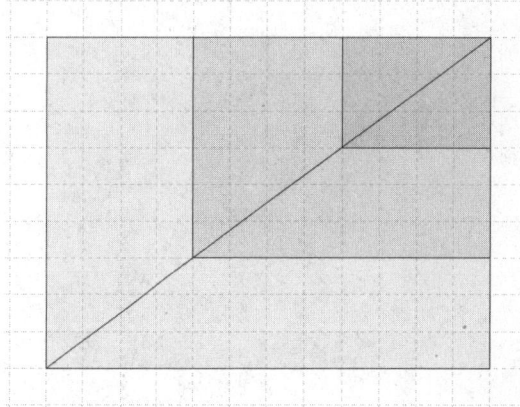

6. A general hexagonal pyramid has height 10 in. A slice 2 in. above the base has area 16 in^2. Find the area of the base.

7. A general cone has base area 3 units^2. Find the area of the slice of the cone that is parallel to the base and $\frac{2}{3}$ of the way from the vertex to the base.

8. A rectangular cone and a triangular cone have bases with the same area. Explain why the cross-sections for the cones halfway between the base and the vertex have the same area.

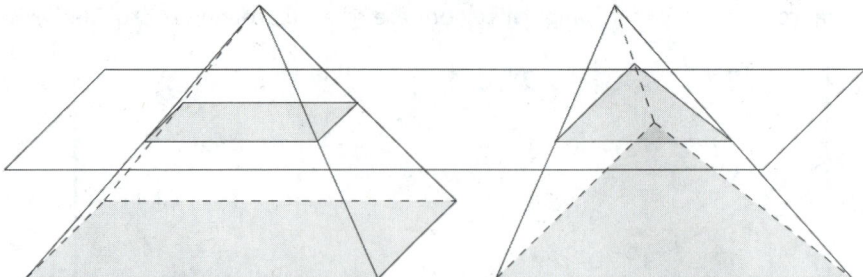

9. The following right triangle is rotated about side AB. What is the resulting figure, and what are its dimensions?

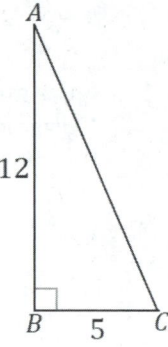

EUREKA
MATH™

Lesson 8: Definition and Properties of Volume

Opening Exercise

a.   Use the following image to reason why the area of a right triangle is $\frac{1}{2}bh$ (Area Property 2).

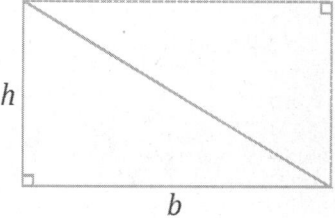

b.   Use the following image to reason why the volume of the following triangular prism with base area A and height h is Ah (Volume Property 2).

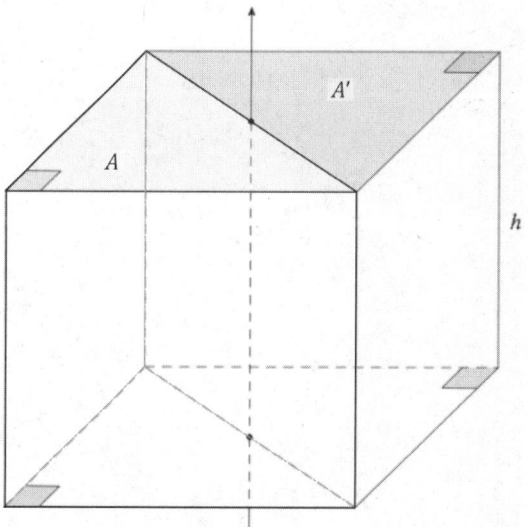

Exercises

Complete Exercises 1–2, and then have a partner check your work.

1. Divide the following polygonal region into triangles. Assign base and height values of your choice to each triangle, and determine the area for the entire polygon.

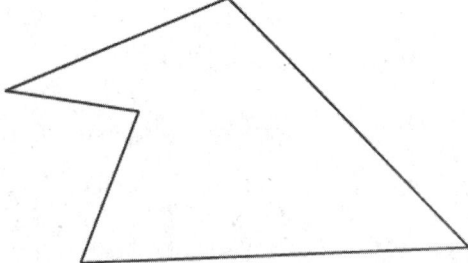

2. The polygon from Exercise 1 is used here as the base of a general right prism. Use a height of 10 and the appropriate value(s) from Exercise 1 to determine the volume of the prism.

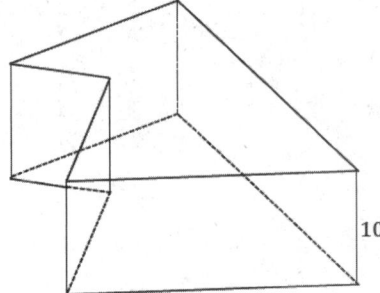

EUREKA
MATH™

© 2015 Great Minds. eureka-math.org
GEO-M3-SE-B2-1.3.0-10.2015

We can use the formula density $= \frac{\text{mass}}{\text{volume}}$ to find the density of a substance.

3. A square metal plate has a density of 10.2 g/cm³ and weighs 2.193 kg.

    a. Calculate the volume of the plate.

    b. If the base of this plate has an area of 25 cm², determine its thickness.

4. A metal cup full of water has a mass of 1,000 g. The cup itself has a mass of 214.6 g. If the cup has both a diameter and a height of 10 cm, what is the approximate density of water?

EUREKA
MATH™

Lesson 8:   Definition and Properties of Volume

S.59

© 2015 Great Minds. eureka-math.org
GEO-M3-SE-B2-1.3.0-10.2015

Problem Set

1. Two congruent solids S_1 and S_2 have the property that $S_1 \cap S_2$ is a right triangular prism with height $\sqrt{3}$ and a base that is an equilateral triangle of side length 2. If the volume of $S_1 \cup S_2$ is 25 units^3, find the volume of S_1.

2. Find the volume of a triangle with side lengths 3, 4, and 5.

3. The base of the prism shown in the diagram consists of overlapping congruent equilateral triangles ABC and DGH. Points C, D, E, and F are midpoints of the sides of triangles ABC and DGH. $GH = AB = 4$, and the height of the prism is 7. Find the volume of the prism.

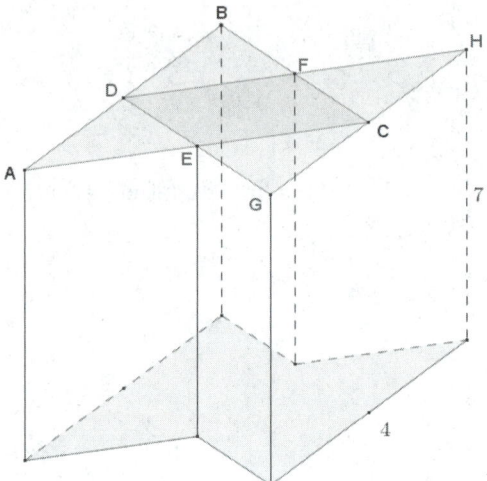

4. Find the volume of a right rectangular pyramid whose base is a square with side length 2 and whose height is 1.

   Hint: Six such pyramids can be fit together to make a cube with side length 2, as shown in the diagram.

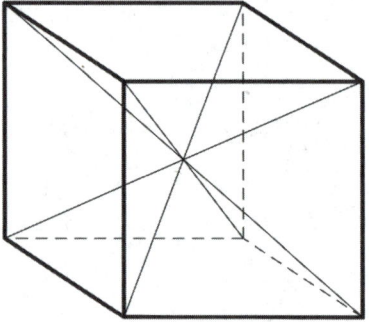

5. Draw a rectangular prism with a square base such that the pyramid's vertex lies on a line perpendicular to the base of the prism through one of the four vertices of the square base, and the distance from the vertex to the base plane is equal to the side length of the square base.

EUREKA
MATH™

6. The pyramid that you drew in Problem 5 can be pieced together with two other identical rectangular pyramids to form a cube. If the side lengths of the square base are 3, find the volume of the pyramid.

7. Paul is designing a mold for a concrete block to be used in a custom landscaping project. The block is shown in the diagram with its corresponding dimensions and consists of two intersecting rectangular prisms. Find the volume of mixed concrete, in cubic feet, needed to make Paul's custom block.

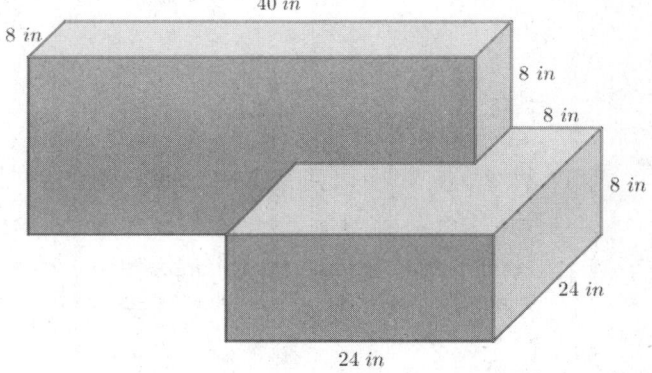

8. Challenge: Use card stock and tape to construct three identical polyhedron nets that together form a cube.

Opening

Area Properties	Volume Properties
1. The area of a set in two dimensions is a number greater than or equal to zero that measures the size of the set and not the shape.	1.
2. The area of a rectangle is given by the formula length × width. The area of a triangle is given by the formula $\frac{1}{2}$ × base × height. A polygonal region is the union of finitely many non-overlapping triangular regions and has area the sum of the areas of the triangles.	2.
3. Congruent regions have the same area.	3. Congruent solids have the same volume.

EUREKA
MATH

4. The area of the union of two regions is the sum of the areas minus the area of the intersection:

$$\text{Area}(A \cup B) = \text{Area}(A) + \text{Area}(B) - \text{Area}(A \cap B).$$

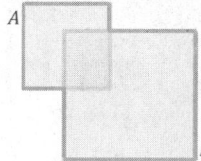

4.

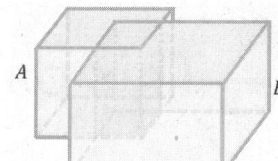

5. The area of the difference of two regions where one is contained in the other is the difference of the areas:

If $A \subseteq B$, then $\text{Area}(B - A) = \text{Area}(B) - \text{Area}(A)$.

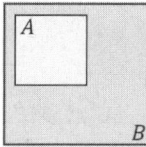

5.

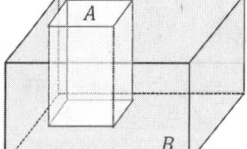

6. The area a of a region A can be estimated by using polygonal regions S and T so that S is contained in A, and A is contained in T.

Then, $\text{Area}(S) \leq a \leq \text{Area}(T)$.

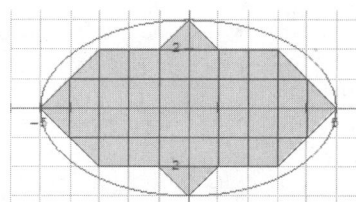

6.

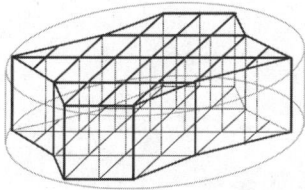

This page intentionally left blank

Lesson 9: Scaling Principle for Volumes

Classwork

Opening Exercise

a. For each pair of similar figures, write the ratio of side lengths $a:b$ or $c:d$ that compares one pair of corresponding sides. Then, complete the third column by writing the ratio that compares the areas of the similar figures. Simplify ratios when possible.

Similar Figures	Ratio of Side Lengths $a:b$ or $c:d$	Ratio of Areas Area(A): Area(B) or Area(C): Area(D)
$\triangle A \sim \triangle B$	6:4 3:2	9:4 $3^2:2^2$
Rectangle $A \sim$ Rectangle B		
$\triangle C \sim \triangle D$		

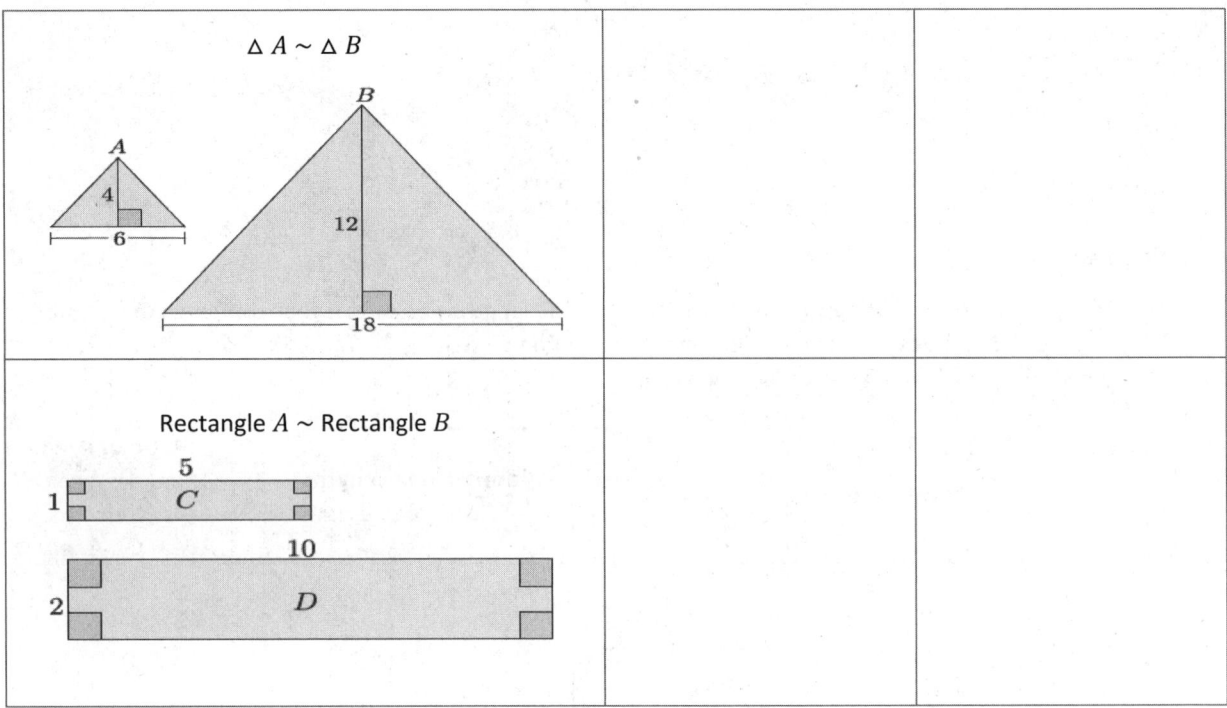

b.

    i.   State the relationship between the ratio of sides $a:b$ and the ratio of the areas $\text{Area}(A):\text{Area}(B)$.

    ii.   Make a conjecture as to how the ratio of sides $a:b$ will be related to the ratio of volumes $\text{Volume}(S):\text{Volume}(T)$. Explain.

c.   What does it mean for two solids in three-dimensional space to be similar?

Lesson 9:   Scaling Principle for Volumes

© 2015 Great Minds. eureka-math.org
GEO-M3-SE-B2-1.3.0-10.2015

Exercises

1. Each pair of solids shown below is similar. Write the ratio of side lengths $a : b$ comparing one pair of corresponding sides. Then, complete the third column by writing the ratio that compares volumes of the similar figures. Simplify ratios when possible.

Similar Figures	Ratio of Side Lengths $a : b$	Ratio of Volumes $\text{Volume}(A) : \text{Volume}(B)$
Figure A Figure B		
Figure A Figure B		
Figure A Figure B		

EUREKA
MATH™

Lesson 9:   Scaling Principle for Volumes

S.67

© 2015 Great Minds. eureka-math.org
GEO-M3-SE-B2-1.3.0-10.2015

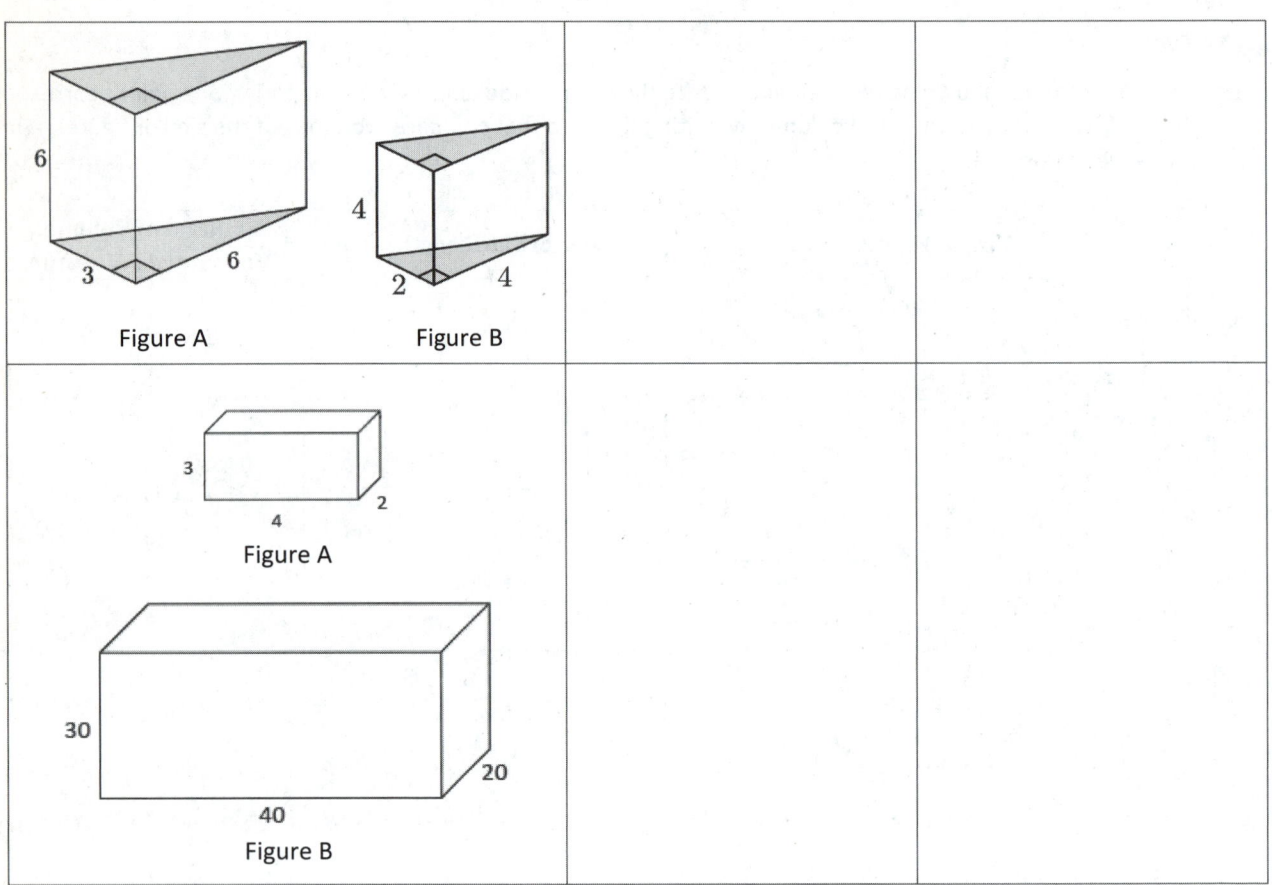

2. Use the triangular prism shown below to answer the questions that follow.

   a. Calculate the volume of the triangular prism.

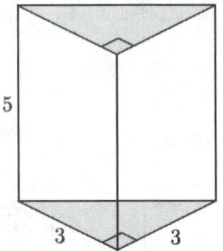

   b. If one side of the triangular base is scaled by a factor of 2, the other side of the triangular base is scaled by a factor of 4, and the height of the prism is scaled by a factor of 3, what are the dimensions of the scaled triangular prism?

EUREKA
MATH™

c.   Calculate the volume of the scaled triangular prism.

d.   Make a conjecture about the relationship between the volume of the original triangular prism and the scaled triangular prism.

e.   Do the volumes of the figures have the same relationship as was shown in the figures in Exercise 1? Explain.

3.   Use the rectangular prism shown to answer the questions that follow.

a.   Calculate the volume of the rectangular prism.

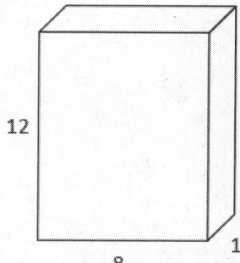

12

8

1

EUREKA
MATH™

Lesson 9:   Scaling Principle for Volumes

S.69

© 2015 Great Minds. eureka-math.org
GEO-M3-SE-B2-1.3.0-10.2015

b. If one side of the rectangular base is scaled by a factor of $\frac{1}{2}$, the other side of the rectangular base is scaled by a factor of 24, and the height of the prism is scaled by a factor of $\frac{1}{3}$, what are the dimensions of the scaled rectangular prism?

c. Calculate the volume of the scaled rectangular prism.

d. Make a conjecture about the relationship between the volume of the original rectangular prism and the scaled rectangular prism.

4. A manufacturing company needs boxes to ship their newest widget, which measures $2 \times 4 \times 5 \text{ in}^3$. Standard size boxes, 5-inch cubes, are inexpensive but require foam packaging so the widget is not damaged in transit. Foam packaging costs $0.03 per cubic inch. Specially designed boxes are more expensive but do not require foam packing. If the standard size box costs $0.80 each, and the specially designed box costs $3.00 each, which kind of box should the company choose? Explain your answer.

Problem Set

1. Coffees sold at a deli come in similar-shaped cups. A small cup has a height of 4.2", and a large cup has a height of 5". The large coffee holds 12 fluid ounces. How much coffee is in a small cup? Round your answer to the nearest tenth of an ounce.

2. Right circular cylinder A has volume 2,700 and radius 3. Right circular cylinder B is similar to cylinder A and has volume 6,400. Find the radius of cylinder B.

3. The Empire State Building is a 102-story skyscraper. Its height is 1,250 ft. from the ground to the roof. The length and width of the building are approximately 424 ft. and 187 ft., respectively. A manufacturing company plans to make a miniature version of the building and sell cases of them to souvenir shops.

   a. The miniature version is just $\dfrac{1}{2500}$ of the size of the original. What are the dimensions of the miniature Empire State Building?

   b. Determine the volume of the minature building. Explain how you determined the volume.

4. If a right square pyramid has a 2 × 2 square base and height 1, then its volume is $\dfrac{4}{3}$. Use this information to find the volume of a right rectangular prism with base dimensions $a \times b$ and height h.

5. The following solids are similar. The volume of the first solid is 100. Find the volume of the second solid.

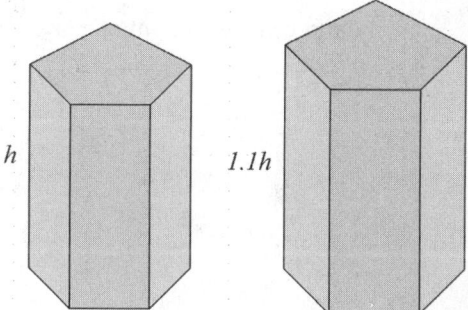

6.   A general cone has a height of 6. What fraction of the cone's volume is between a plane containing the base and a parallel plane halfway between the vertex of the cone and the base plane?

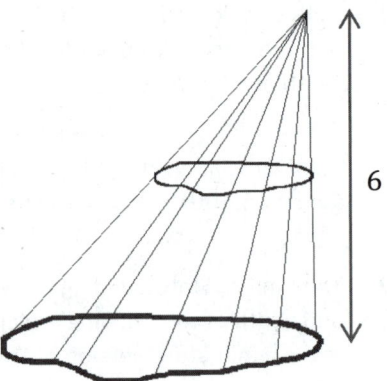

7.   A company uses rectangular boxes to package small electronic components for shipping. The box that is currently used can contain 500 of one type of component. The company wants to package twice as many pieces per box. Michael thinks that the box will hold twice as much if its dimensions are doubled. Shawn disagrees and says that Michael's idea provides a box that is much too large for 1,000 pieces. Explain why you agree or disagree with one or either of the boys. What would you recommend to the company?

8.   A dairy facility has bulk milk tanks that are shaped like right circular cylinders. They have replaced one of their bulk milk tanks with three smaller tanks that have the same height as the original but $\frac{1}{3}$ the radius. Do the new tanks hold the same amount of milk as the original tank? If not, explain how the volumes compare.

EUREKA
MATH™

Lesson 10: The Volume of Prisms and Cylinders and Cavalieri's Principle

Opening Exercise

The bases of the following triangular prism T and rectangular prism R lie in the same plane. A plane that is parallel to the bases and also a distance 3 from the bottom base intersects both solids and creates cross-sections T' and R'.

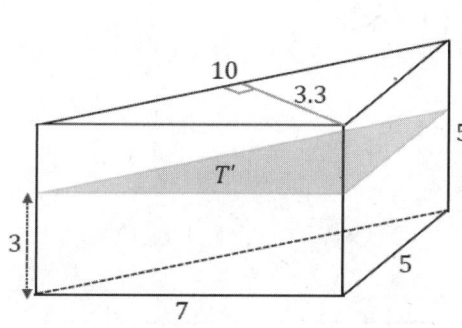 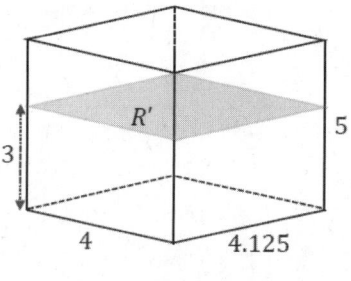

a.   Find Area(T').

b.   Find Area(R').

c.   Find Vol(T).

d.   Find Vol(R).

e. If a height other than 3 were chosen for the cross-section, would the cross-sectional area of either solid change?

Discussion

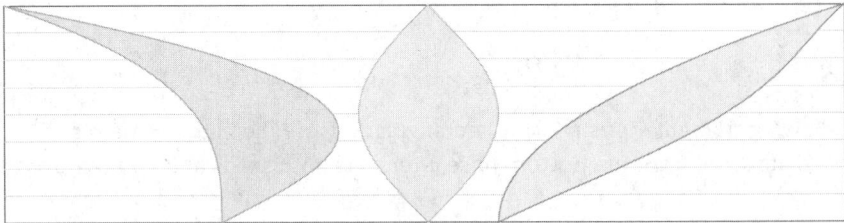

Figure 1

Example 1

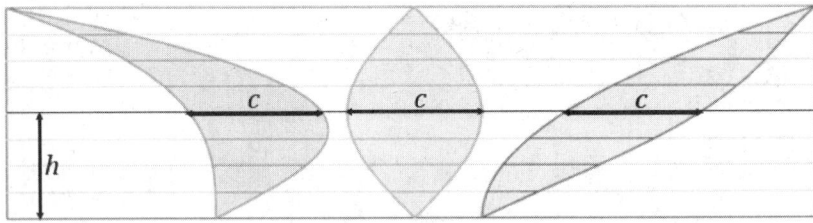

Example 2

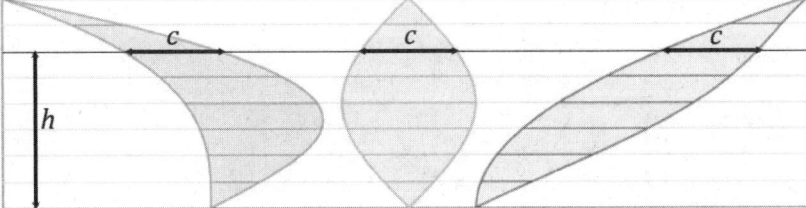

PRINCIPLE OF PARALLEL SLICES IN THE PLANE: If two planar figures of equal altitude have identical cross-sectional lengths at each height, then the regions of the figures have the same area.

EUREKA
MATH™

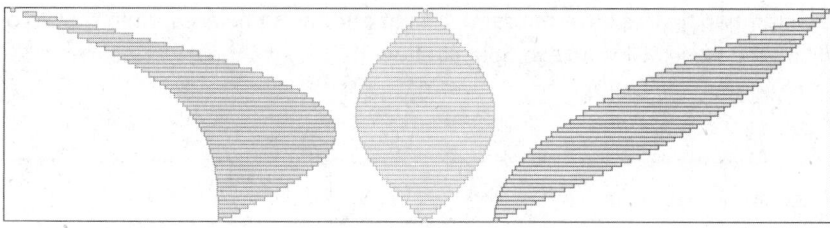

Figure 2

Example

a. The following triangles have equal areas: $\text{Area}(\triangle ABC) = \text{Area}(\triangle A'B'C') = 15$ units^2. The distance between $\overrightarrow{DE}$ and $\overleftrightarrow{CC'}$ is 3. Find the lengths of $\overline{DE}$ and $\overline{D'E'}$.

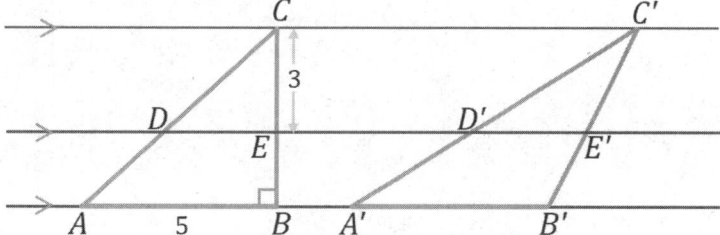

EUREKA
MATH™

© 2015 Great Minds. eureka-math.org
GEO-M3-SE-B2-1.3.0-10.2015

b. Joey says that if two figures have the same height and the same area, then their cross-sectional lengths at each height will be the same. Give an example to show that Joey's theory is incorrect.

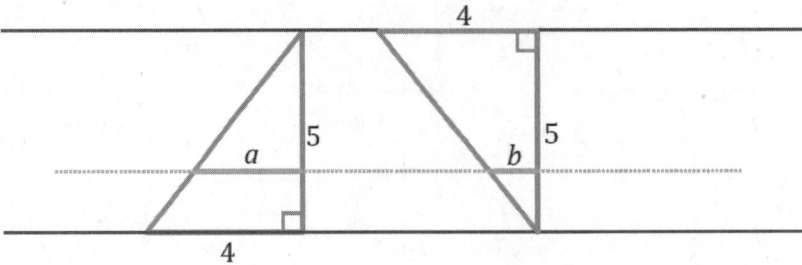

Discussion

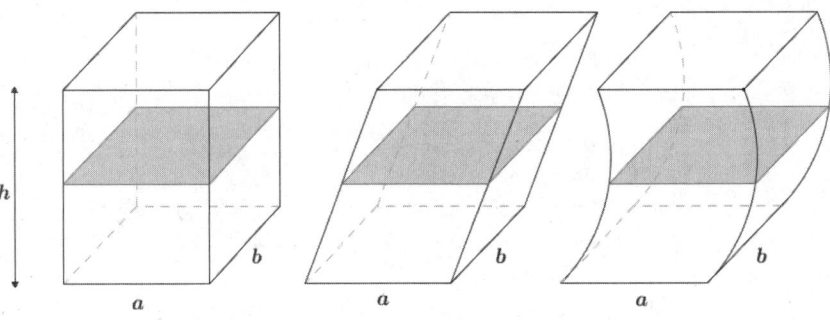

Figure 3

CAVALIERI'S PRINCIPLE: Given two solids that are included between two parallel planes, if every plane parallel to the two planes intersects both solids in cross-sections of equal area, then the volumes of the two solids are equal.

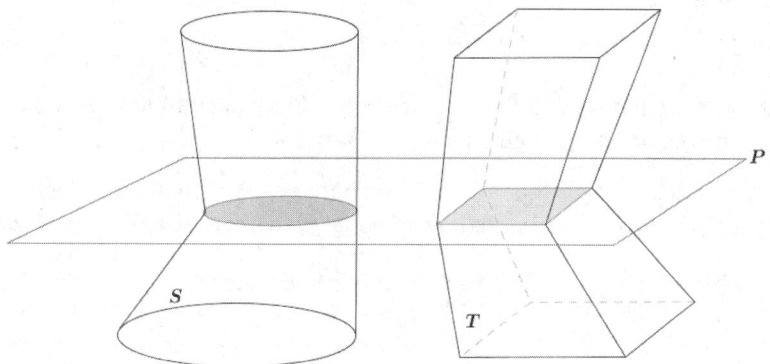

Figure 4

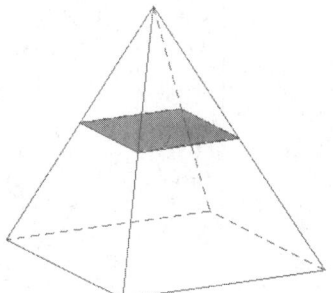
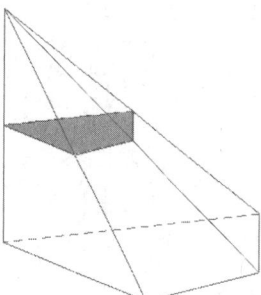

Figure 5

Figure 6

> **Lesson Summary**
>
> **PRINCIPLE OF PARALLEL SLICES IN THE PLANE:** If two planar figures of equal altitude have identical cross-sectional lengths at each height, then the regions of the figures have the same area.
>
> **CAVALIERI'S PRINCIPLE:** Given two solids that are included between two parallel planes, if every plane parallel to the two planes intersects both solids in cross-sections of equal area, then the volumes of the two solids are equal.

Problem Set

1. Use the principle of parallel slices to explain the area formula for a parallelogram.

2. Use the principle of parallel slices to show that the three triangles shown below all have the same area.

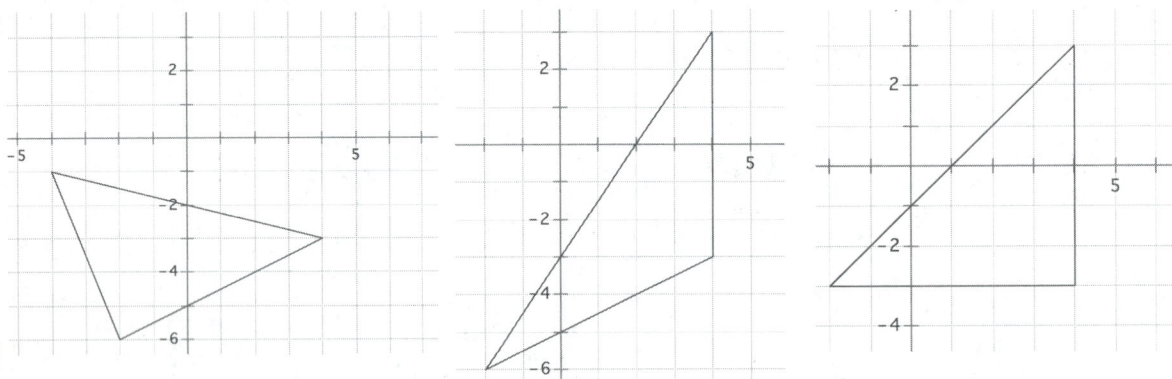

Figure 1   Figure 2   Figure 3

3. An oblique prism has a rectangular base that is 16 in. × 9 in. A hole in the prism is also the shape of an oblique prism with a rectangular base that is 3 in. wide and 6 in. long, and the prism's height is 9 in. (as shown in the diagram). Find the volume of the remaining solid.

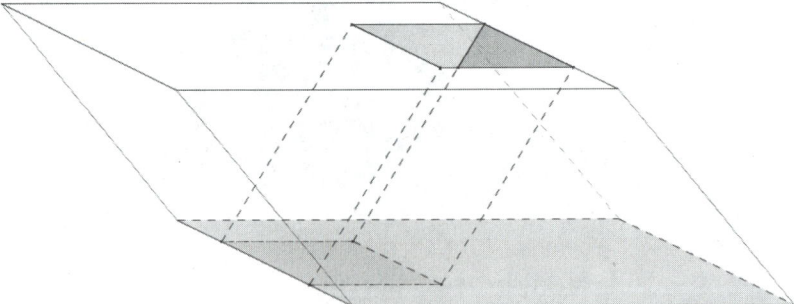

EUREKA
MATH™

4. An oblique circular cylinder has height 5 and volume 45π. Find the radius of the circular base.

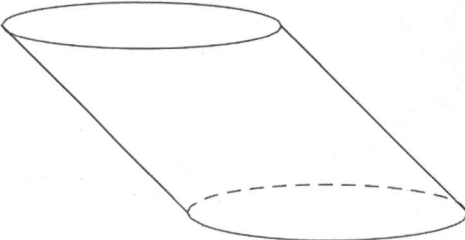

5. A right circular cone and a solid hemisphere share the same base. The vertex of the cone lies on the hemisphere. Removing the cone from the solid hemisphere forms a solid. Draw a picture, and describe the cross-sections of this solid that are parallel to the base.

6. Use Cavalieri's principle to explain why a circular cylinder with a base of radius 5 and a height of 10 has the same volume as a square prism whose base is a square with edge length $5\sqrt{\pi}$ and whose height is also 10.

© 2015 Great Minds. eureka-math.org
GEO-M3-SE-B2-1.3.0-10.2015

This page intentionally left blank

Lesson 11: The Volume Formula of a Pyramid and Cone

Classwork

Exploratory Challenge

Use the provided manipulatives to aid you in answering the questions below.

a.

    i.   What is the formula to find the area of a triangle?

    ii.   Explain why the formula works.

b.

    i.   What is the formula to find the volume of a triangular prism?

    ii.   Explain why the formula works.

c.

    i.   What is the formula to find the volume of a cone or pyramid?

    ii.   Explain why the formula works.

Exercises

1. A cone fits inside a cylinder so that their bases are the same and their heights are the same, as shown in the diagram below. Calculate the volume that is inside the cylinder but outside of the cone. Give an exact answer.

    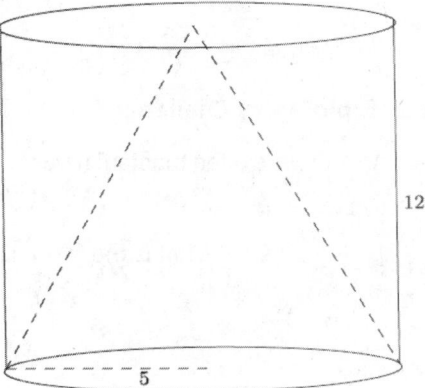

2. A square pyramid has a volume of 245 in³. The height of the pyramid is 15 in. What is the area of the base of the pyramid? What is the length of one side of the base?

EUREKA
MATH™

3. Use the diagram below to answer the questions that follow.

    a. Determine the volume of the cone shown below. Give an exact answer.

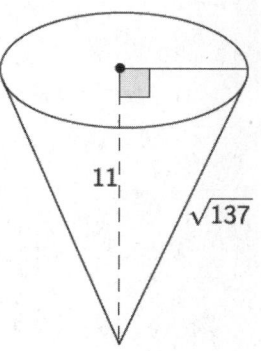

    b. Find the dimensions of a cone that is similar to the one given above. Explain how you found your answers.

    c. Calculate the volume of the cone that you described in part (b) in two ways. (Hint: Use the volume formula and the scaling principle for volume.)

4. Gold has a density of 19.32 g/cm^3. If a square pyramid has a base edge length of 5 cm, height of 6 cm, and a mass of 942 g, is the pyramid in fact solid gold? If it is not, what reasons could explain why it is not? Recall that density can be calculated with the formula density $= \dfrac{\text{mass}}{\text{volume}}$.

EUREKA
MATH™

Problem Set

1. What is the volume formula for a right circular cone with radius r and height h?

2. Identify the solid shown, and find its volume.

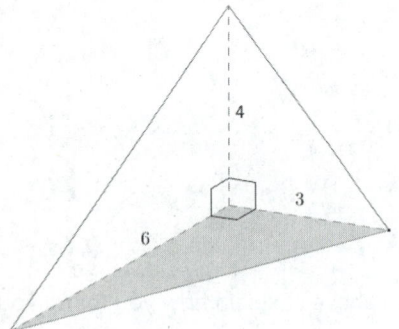

3. Find the volume of the right rectangular pyramid shown.

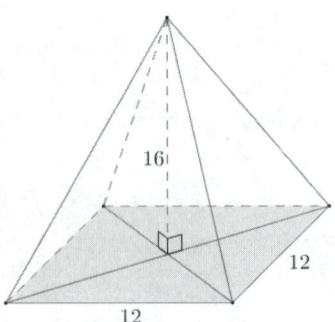

4. Find the volume of the circular cone in the diagram. (Use $\frac{22}{7}$ as an approximation of pi.)

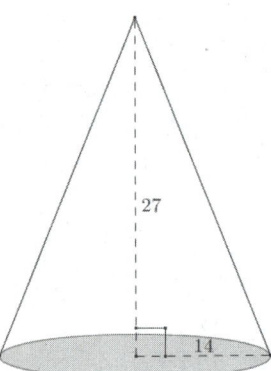

EUREKA
MATH™

5. Find the volume of a pyramid whose base is a square with edge length 3 and whose height is also 3.

6. Suppose you fill a conical paper cup with a height of 6" with water. If all the water is then poured into a cylindrical cup with the same radius and same height as the conical paper cup, to what height will the water reach in the cylindrical cup?

7. Sand falls from a conveyor belt and forms a pile on a flat surface. The diameter of the pile is approximately 10 ft., and the height is approximately 6 ft. Estimate the volume of the pile of sand. State your assumptions used in modeling.

8. A pyramid has volume 24 and height 6. Find the area of its base.

9. Two jars of peanut butter by the same brand are sold in a grocery store. The first jar is twice the height of the second jar, but its diameter is one-half as much as the shorter jar. The taller jar costs $1.49, and the shorter jar costs $2.95. Which jar is the better buy?

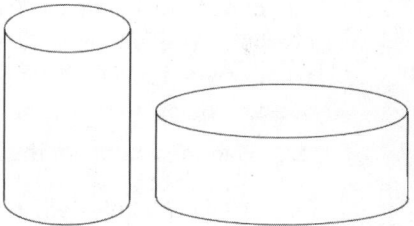

10. A cone with base area A and height h is sliced by planes parallel to its base into three pieces of equal height. Find the volume of each section.

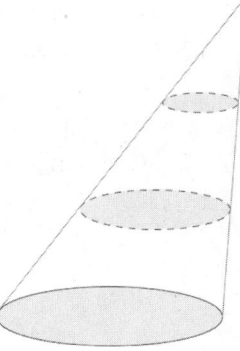

EUREKA
MATH™

Lesson 11:   The Volume Formula of a Pyramid and Cone

S.85

© 2015 Great Minds. eureka-math.org
GEO-M3-SE-B2-1.3.0-10.2015

11. The frustum of a pyramid is formed by cutting off the top part by a plane parallel to the base. The base of the pyramid and the cross-section where the cut is made are called the *bases of the frustum*. The distance between the planes containing the bases is called the *height of the frustum*. Find the volume of a frustum if the bases are squares of edge lengths 2 and 3, and the height of the frustum is 4.

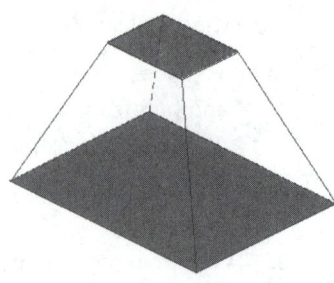

12. A bulk tank contains a heavy grade of oil that is to be emptied from a valve into smaller 5.2-quart containers via a funnel. To improve the efficiency of this transfer process, Jason wants to know the greatest rate of oil flow that he can use so that the container and funnel do not overflow. The funnel consists of a cone that empties into a circular cylinder with the dimensions as shown in the diagram. Answer each question below to help Jason determine a solution to his problem.

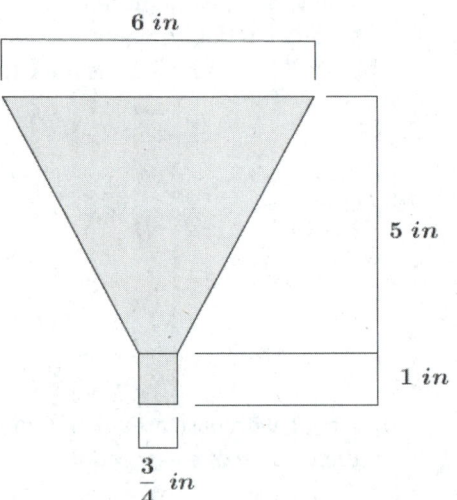

a. Find the volume of the funnel.

b. If 1 in^3 is equivalent in volume to $\dfrac{4}{231}$ qt., what is the volume of the funnel in quarts?

c. If this particular grade of oil flows out of the funnel at a rate of 1.4 quarts per minute, how much time in minutes is needed to fill the 5.2-quart container?

d. Will the tank valve be shut off exactly when the container is full? Explain.

e. How long after opening the tank valve should Jason shut the valve off?

f. What is the maximum constant rate of flow from the tank valve that will fill the container without overflowing either the container or the funnel?

Lesson 11:   The Volume Formula of a Pyramid and Cone

Lesson 12: The Volume Formula of a Sphere

Classwork

Opening Exercise

Picture a marble and a beach ball. Which one would you describe as a sphere? What differences between the two could possibly impact how we describe what a sphere is?

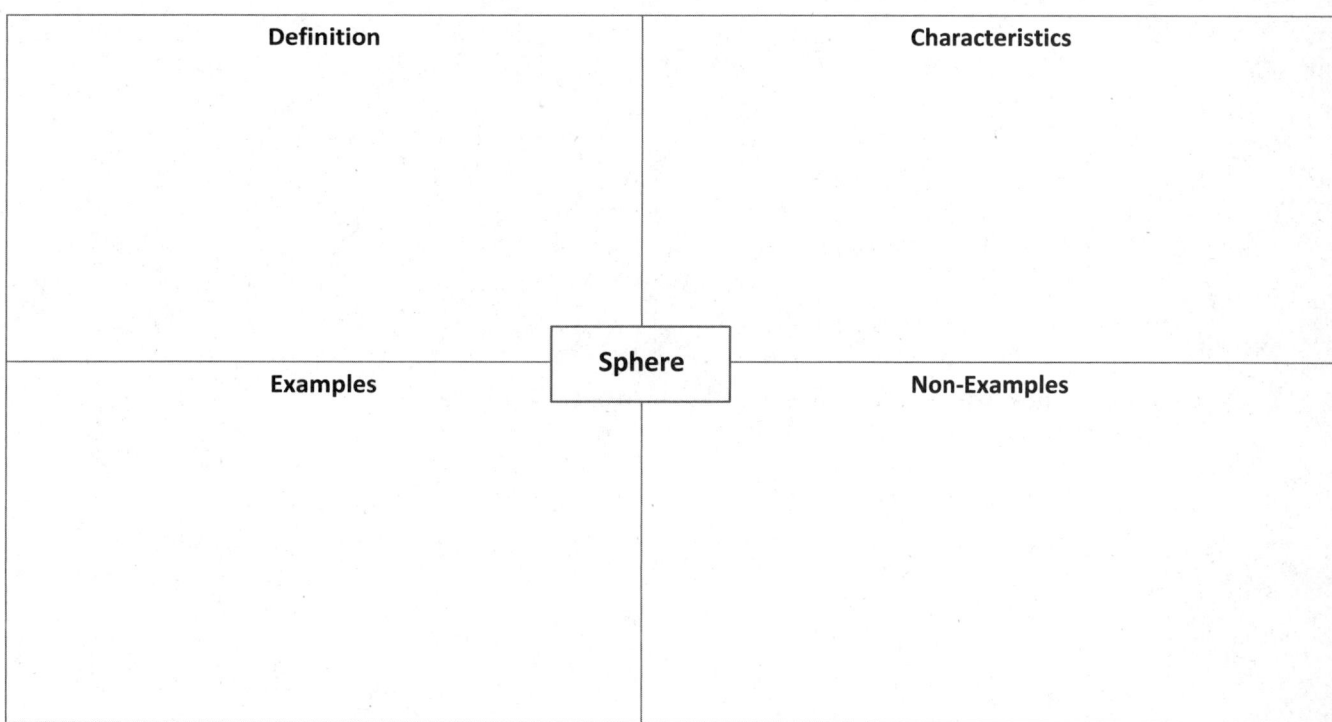

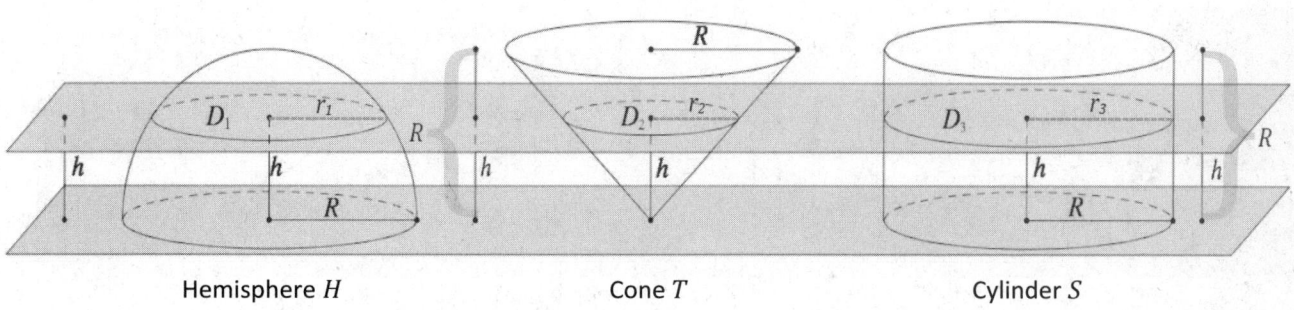

Hemisphere H Cone T Cylinder S

Example

Use your knowledge about the volumes of cones and cylinders to find a volume for a solid hemisphere of radius R.

Exercises

1. Find the volume of a sphere with a diameter of 12 cm to one decimal place.

2. An ice cream cone is 11 cm deep and 5 cm across the opening of the cone. Two hemisphere-shaped scoops of ice cream, which also have diameters of 5 cm, are placed on top of the cone. If the ice cream were to melt into the cone, would it overflow?

3. Bouncy rubber balls are composed of a hollow rubber shell 0.4" thick and an outside diameter of 1.2". The price of the rubber needed to produce this toy is $0.035/\text{in}^3$.

   a. What is the cost of producing 1 case, which holds 50 such balls? Round to the nearest cent.

   b. If each ball is sold for $0.10, how much profit is earned on each ball sold?

Extension

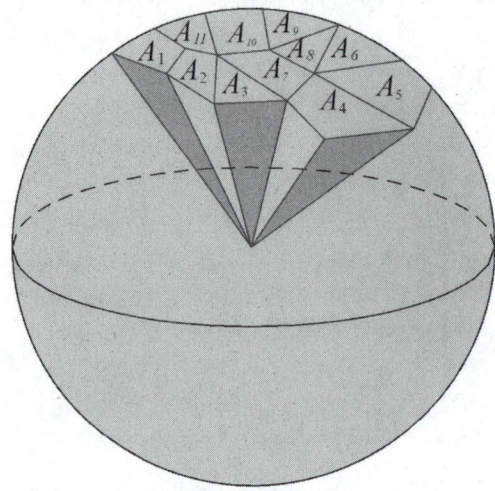

EUREKA
MATH™

Lesson Summary

SPHERE: Given a point C in the three-dimensional space and a number $r > 0$, the *sphere with center C and radius r* is the set of all points in space that are distance r from the point C.

SOLID SPHERE OR BALL: Given a point C in the three-dimensional space and a number $r > 0$, the *solid sphere (or ball)* *with center C and radius r* is the set of all points in space whose distance from the point C is less than or equal to r.

Problem Set

1. A solid sphere has volume 36π. Find the radius of the sphere.

2. A sphere has surface area 16π. Find the radius of the sphere.

3. Consider a right circular cylinder with radius r and height h. The area of each base is πr^2. Think of the lateral surface area as a label on a soup can. If you make a vertical cut along the label and unroll it, the label unrolls to the shape of a rectangle.

    a. Find the dimensions of the rectangle.

    b. What is the lateral (or curved) area of the cylinder?

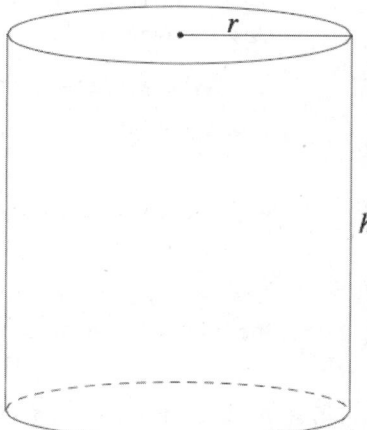

© 2015 Great Minds. eureka-math.org
GEO-M3-SE-B2-1.3.0-10.2015

4. Consider a right circular cone with radius r, height h, and slant height l (see Figure 1). The area of the base is πr^2. Open the lateral area of the cone to form part of a disk (see Figure 2). The surface area is a fraction of the area of this disk.

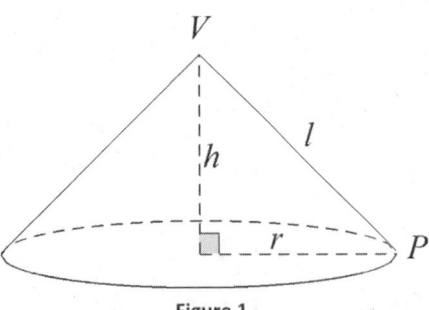

Figure 1

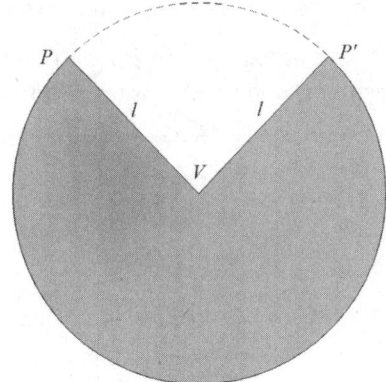

Figure 2

a. What is the area of the entire disk in Figure 2?

b. What is the circumference of the disk in Figure 2?

The length of the arc on this circumference (i.e., the arc that borders the green region) is the circumference of the base of the cone with radius r or $2\pi r$. (Remember, the green region forms the curved portion of the cone and closes around the circle of the base.)

c. What is the ratio of the area of the disk that is shaded to the area of the whole disk?

d. What is the lateral (or curved) area of the cone?

5. A right circular cone has radius 3 cm and height 4 cm. Find the lateral surface area.

6. A semicircular disk of radius 3 ft. is revolved about its diameter (straight side) one complete revolution. Describe the solid determined by this revolution, and then find the volume of the solid.

7. A sphere and a circular cylinder have the same radius, r, and the height of the cylinder is $2r$.

a. What is the ratio of the volumes of the solids?

b. What is the ratio of the surface areas of the solids?

8. The base of a circular cone has a diameter of 10 cm and an altitude of 10 cm. The cone is filled with water. A sphere is lowered into the cone until it just fits. Exactly one-half of the sphere remains out of the water. Once the sphere is removed, how much water remains in the cone?

Lesson 12:   The Volume Formula of a Sphere

© 2015 Great Minds. eureka-math.org
GEO-M3-SE-B2-1.3.0-10.2015

EUREKA
MATH™

9. Teri has an aquarium that is a cube with edge lengths of 24 inches. The aquarium is $\frac{2}{3}$ full of water. She has a supply of ball bearings, each having a diameter of $\frac{3}{4}$ inch.

    a. What is the maximum number of ball bearings that Teri can drop into the aquarium without the water overflowing?

    b. Would your answer be the same if the aquarium were $\frac{2}{3}$ full of sand? Explain.

    c. If the aquarium were empty, how many ball bearings would fit on the bottom of the aquarium if you arranged them in rows and columns, as shown in the picture?

    d. How many of these layers could be stacked inside the aquarium without going over the top of the aquarium? How many bearings would there be altogether?

    e. With the bearings still in the aquarium, how much water can be poured into the aquarium without overflowing?

    f. Approximately how much of the aquarium do the ball bearings occupy?

10. Challenge: A hemispherical bowl has a radius of 2 meters. The bowl is filled with water to a depth of 1 meter. What is the volume of water in the bowl? (Hint: Consider a cone with the same base radius and height and the cross-section of that cone that lies 1 meter from the vertex.)

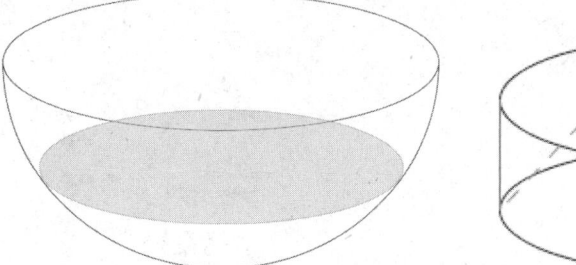
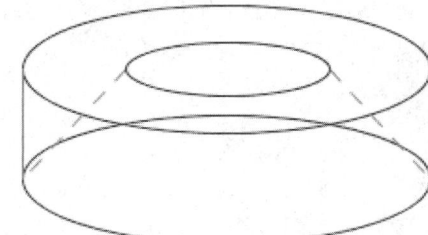

11. Challenge: A certain device must be created to house a scientific instrument. The housing must be a spherical shell, with an outside diameter of 1 m. It will be made of a material whose density is 14 g/cm^3. It will house a sensor inside that weighs 1.2 kg. The housing, with the sensor inside, must be neutrally buoyant, meaning that its density must be the same as water. Ignoring any air inside the housing, and assuming that water has a density of 1 g/cm^3, how thick should the housing be made so that the device is neutrally buoyant? Round your answer to the nearest tenth of a centimeter.

12. Challenge: An inverted, conical tank has a circular base of radius 2 m and a height of 2 m and is full of water. Some of the water drains into a hemispherical tank, which also has a radius of 2 m. Afterward, the depth of the water in the conical tank is 80 cm. Find the depth of the water in the hemispherical tank.

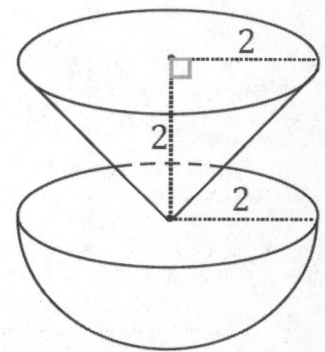

This page intentionally left blank

Lesson 13: How Do 3D Printers Work?

Opening Exercise

a. Observe the following right circular cone. The base of the cone lies in plane S, and planes P, Q, and R are all parallel to S. Plane P contains the vertex of the cone.

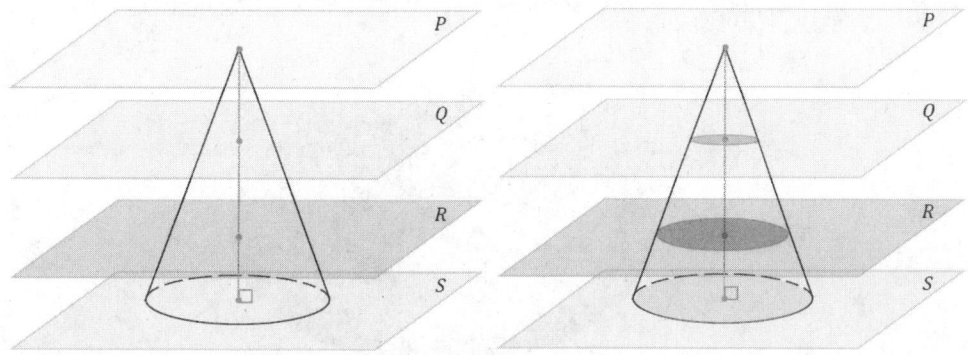

Sketch the cross-section P' of the cone by plane P.	Sketch the cross-section Q' of the cone by plane Q.
Sketch the cross-section R' of the cone by plane R.	Sketch the cross-section S' of the cone by plane S.

b.   What happens to the cross-sections as we look at them starting with P' and work toward S'?

Exercise 1

1.   Sketch five evenly spaced, horizontal cross-sections made with the following figure.

http://commons.wikimedia.org/wiki/File%3ATorus_illustration.png; By Oleg Alexandrov (self-made, with MATLAB) [Public domain], via Wikimedia Commons. Attribution not legally required.

Example

Let us now try drawing cross-sections of an everyday object, such as a coffee cup.

Sketch the cross-sections at each of the indicated heights.

1	2	3

4	5

Exercises 2–4

2. A cone with a radius of 5 cm and height of 8 cm is to be printed from a 3D printer. The medium that the printer will use to print (i.e., the "ink" of this 3D printer) is a type of plastic that comes in coils of tubing that has a radius of $1\frac{1}{3}$ cm. What length of tubing is needed to complete the printing of this cone?

3. A cylindrical dessert 8 cm in diameter is to be created using a type of 3D printer specially designed for gourmet kitchens. The printer will "pipe" or, in other words, "print out" the delicious filling of the dessert as a solid cylinder. Each dessert requires 300 cm^3 of filling. Approximately how many layers does each dessert have if each layer is 3 mm thick?

EUREKA
MATH™

4. The image shown to the right is of a fine tube that is printed from a 3D printer that prints replacement parts. If each layer is 2 mm thick, and the printer prints at a rate of roughly 1 layer in 3 seconds, how many minutes will it take to print the tube?

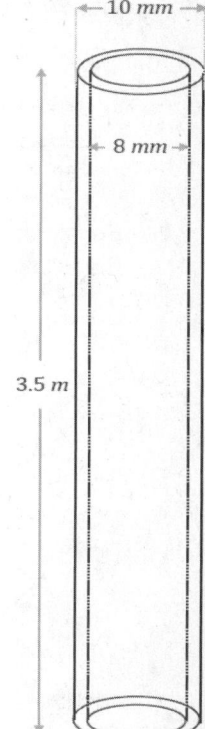

Note: Figure not drawn to scale.

EUREKA
MATH™

Problem Set

1. Horizontal slices of a solid are shown at various levels arranged from highest to lowest. What could the solid be?

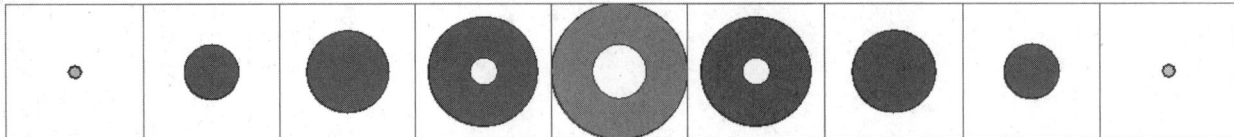

2. Explain the difference in a 3D printing of the ring pictured in Figure 1 and Figure 2 if the ring is oriented in each of the following ways.

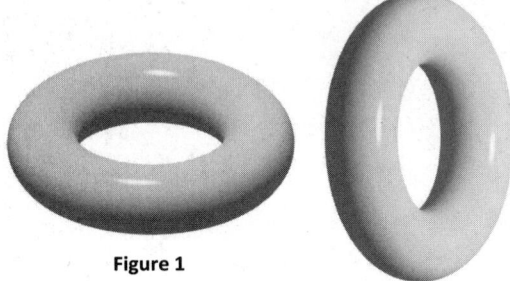

Figure 1

Figure 2

3. Each bangle printed by a 3D printer has a mass of exactly 25 g of metal. If the density of the metal is 14 g/cm^3, what length of a wire 1 mm in radius is needed to produce each bangle? Find your answer to the tenths place.

4. A certain 3D printer uses 100 m of plastic filament that is 1.75 mm in diameter to make a cup. If the filament has a density of 0.32 g/cm^3, find the mass of the cup to the tenths place.

5. When producing a circular cone or a hemisphere with a 3D printer, the radius of each layer of printed material must change in order to form the correct figure. Describe how radius must change in consecutive layers of each figure.

6. Suppose you want to make a 3D printing of a cone. What difference does it make if the vertex is at the top or at the bottom? Assume that the 3D printer places each new layer on top of the previous layer.

7. Filament for 3D printing is sold in spools that contain something shaped like a wire of diameter 3 mm. John wants to make 3D printings of a cone with radius 2 cm and height 3 cm. The length of the filament is 25 meters. About how many cones can John make?

EUREKA
MATH™

8. John has been printing solid cones but would like to be able to produce more cones per each length of filament than you calculated in Problem 7. Without changing the outside dimensions of his cones, what is one way that he could make a length of filament last longer? Sketch a diagram of your idea, and determine how much filament John would save per piece. Then, determine how many cones John could produce from a single length of filament based on your design.

9. A 3D printer uses one spool of filament to produce 20 congruent solids. Suppose you want to produce similar solids that are 10% longer in each dimension. How many such figures could one spool of filament produce?

10. A fabrication company 3D-prints parts shaped like a pyramid with base as shown in the following figure. Each pyramid has a height of 3 cm. The printer uses a wire with a density of 12 g/cm^3 at a cost of $0.07/g.

    It costs $500 to set up for a production run, no matter how many parts they make. If they can only charge $15 per part, how many do they need to make in a production run to turn a profit?

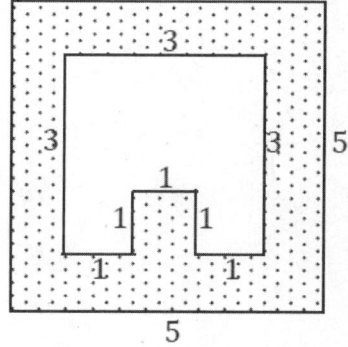

EUREKA
MATH™

© 2015 Great Minds. eureka-math.org
GEO-M3-SE-B2-1.3.0-10.2015

This page intentionally left blank

Student Edition

Eureka Math
Geometry
Module 4

Special thanks go to the Gordon A. Cain Center and to the Department of Mathematics at Louisiana State University for their support in the development of *Eureka Math*.

For a free *Eureka Math* Teacher Resource Pack, Parent Tip Sheets, and more please visit www.Eureka.tools

Printed in the U.S.A.
This book may be purchased from the publisher at eureka-math.org
1 2 3 4 5 6 7 8 BAB 25 24 23 22 21
ISBN 978-1-63255-328-7

Lesson 1: Searching a Region in the Plane

Classwork

Exploratory Challenge

Students in a robotics class must program a robot to move about an empty rectangular warehouse. The program specifies location at a given time t seconds. The room is twice as long as it is wide. Locations are represented as points in a coordinate plane with the southwest corner of the room deemed the origin, $(0,0)$, and the northeast corner deemed the point $(2000,1000)$ in feet, as shown in the diagram below.

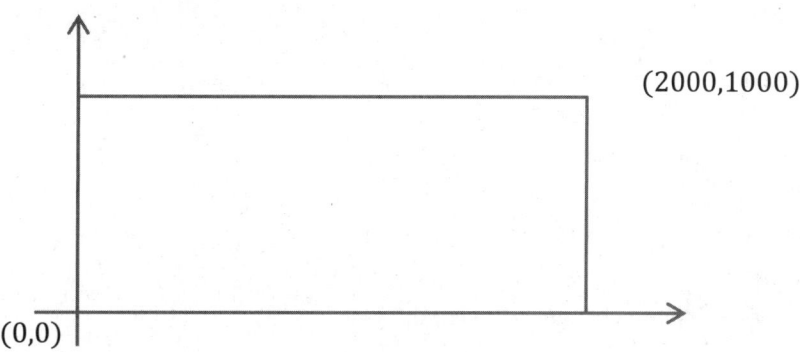

The first program written has the robot moving at a constant speed in a straight line. At time $t = 1$ second, the robot is at position $(30,45)$, and at $t = 3$ seconds, it is at position $(50,75)$. Complete the exercises, and answer the questions below to program the robot's motion.

   a.   At what location will the robot hit the wall?

   b.   At what speed will the robot hit the wall?

   c.   At what time will the robot hit the wall?

Exercises

1. Plot the points on a coordinate plane.

2. Draw the segment connecting the points.

3. How much did the x-coordinate change in 2 seconds?

4. How much did the y-coordinate change in 2 seconds?

5. What is the ratio of change in y to the change in x?

6. What is the equation of the line of motion?

7. What theorem could be used to find the distance between the points?

8. How far did the robot travel in 2 seconds?

EUREKA
MATH™

Problem Set

1. A robot from the video now moves around an empty 100 ft. by 100 ft. storage room at a constant speed. If the robot crosses $(10,10)$ at 1 second and $(30,30)$ at 6 seconds:

   a. Plot the points, and draw the segment connecting the points.

   b. What was the change in the x-coordinate?

   c. What was the change in the y-coordinate?

   d. What is the ratio of the change in y to the change in x?

   e. How far did the robot travel between the two points?

   f. What was the speed of the robot?

   g. Where did the robot start?

2. Your mother received a robot vacuum cleaner as a gift and wants you to help her program it to clean a vacant 30 ft. by 30 ft. room. If the vacuum is at position $(12,9)$ at time $t = 2$ seconds and at position $(24,18)$ at $t = 5$ seconds, answer the following:

   a. How far did the robot travel over 3 seconds?

   b. What is the constant speed of the robot?

   c. What is the ratio of the change in the x-coordinate to the change in the y-coordinate?

   d. Where did the robot start?

   e. Where will the robot be at $t = 3$ seconds? Explain how you know.

   f. At what location will the robot hit the wall?

   g. At what time will the robot hit the wall?

3. A baseball player hits a ball at home plate at position $(0,0)$. It travels at a constant speed across first base at position $(90,0)$ in 2 seconds.

   a. What was the speed of the ball?

   b. When will it cross the fence at position $(300,0)$? Explain how you know.

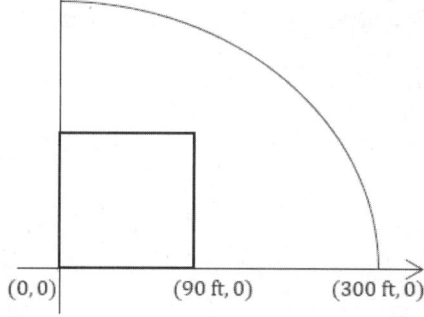

4. The tennis team has a robot that picks up tennis balls. The tennis court is 36 feet wide and 78 feet long. The robot starts at position $(8,10)$ and is at position $(16,20)$ at $t = 4$ seconds after moving at a constant speed. When will it pick up the ball located at position $(28,35)$?

This page intentionally left blank

Lesson 2: Finding Systems of Inequalities That Describe Triangular and Rectangular Regions

Classwork

Opening Exercise

Graph each system of inequalities.

a. $\begin{cases} y \geq 1 \\ x \leq 5 \end{cases}$

    i.   Is $(1,2)$ a solution? Explain.

    ii. Is $(1,1)$ a solution? Explain.

    iii. The region is the intersection of how many half-planes? Explain how you know.

b. $\begin{cases} y < 2x + 1 \\ y \geq -3x - 2 \end{cases}$

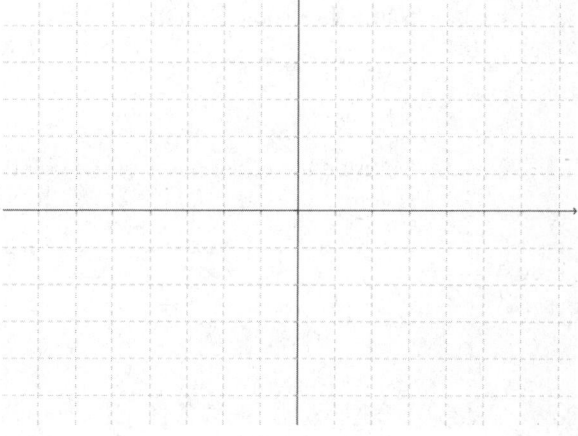

    i.   Is $(-2,4)$ in the solution set?

    ii. Is $(1,3)$ in the solution set?

iii.   The region is the intersection of how many half-planes? Explain how you know.

Example 1

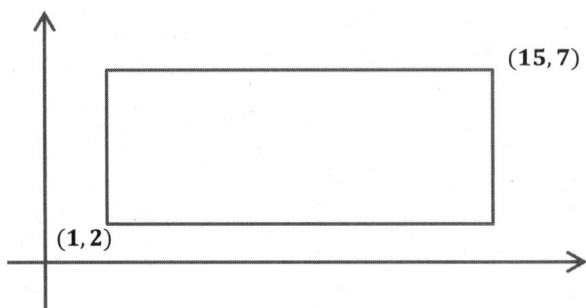

Exercises 1–3

1.   Given the region shown to the right:

   a.   Name three points in the interior of the region.

   b.   Name three points on the boundary.

   c.   Describe the coordinates of the points in the region.

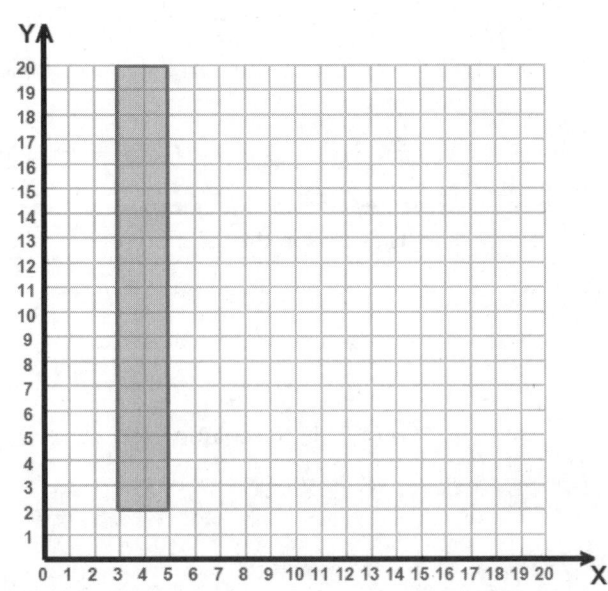

   Lesson 2:   Finding Systems of Inequalities That Describe Triangular and Rectangular Regions

EUREKA MATH™

d.   Write the inequality describing the x-values.

e.   Write the inequality describing the y-values.

f.   Write this as a system of equations.

g.   Will the lines $x = 4$ and $y = 1$ pass through the region? Draw them.

2.   Given the region that continues unbound to the right as shown to the right:

a.   Name three points in the region.

b.   Describe in words the points in the region.

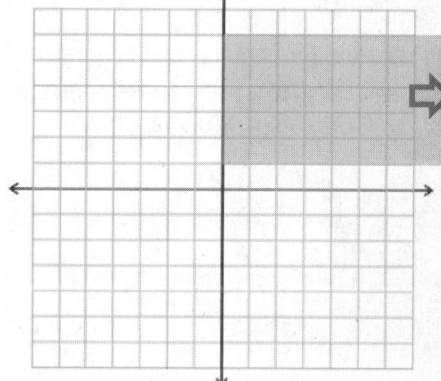

c.   Write the system of inequalities that describe the region.

d.   Name a horizontal line that passes through the region.

EUREKA
MATH™

Lesson 2:   Finding Systems of Inequalities That Describe Triangular and
              Rectangular Regions

S.7

© 2015 Great Minds. eureka-math.org
GEO-M3-SE-B2-1.3.0-10.2015

3. Given the region that continues down without bound as shown to the right:

    a. Describe the region in words.

    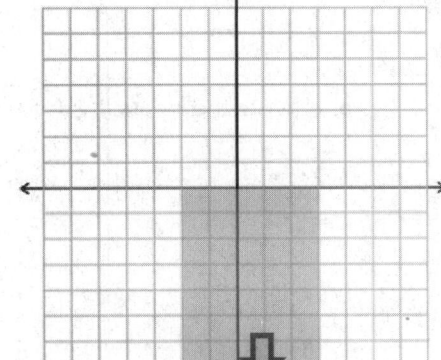

    b. Write the system of inequalities that describe the region.

    c. Name a vertical line that passes through the region.

Example 2

Draw the triangular region in the plane given by the triangle with vertices $(0,0)$, $(1,3)$, and $(2,1)$. Can we write a set of inequalities that describes this region?

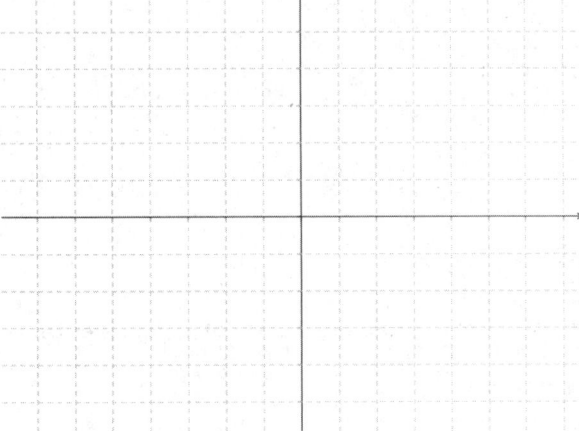

Lesson 2:   Finding Systems of Inequalities That Describe Triangular and
Rectangular Regions

EUREKA
MATH™

Exercises 4–5

4. Given the triangular region shown, describe this region with a system of inequalities.

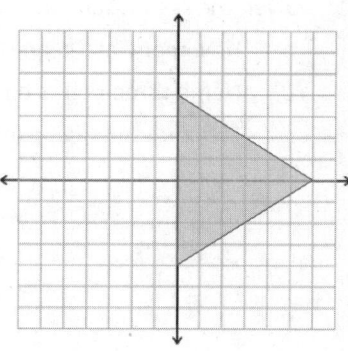

5. Given the trapezoid with vertices $(-2,0)$, $(-1,4)$, $(1,4)$, and $(2,0)$, describe this region with a system of inequalities.

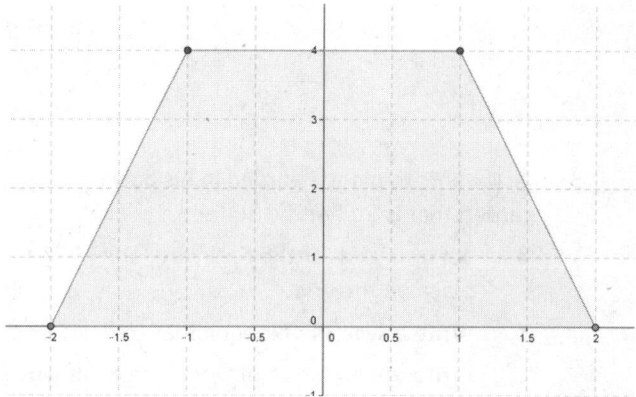

Problem Set

1. Given the region shown:

    a. How many half-planes intersect to form this region?

    b. Name three points on the boundary of the region.

    c. Describe the region in words.

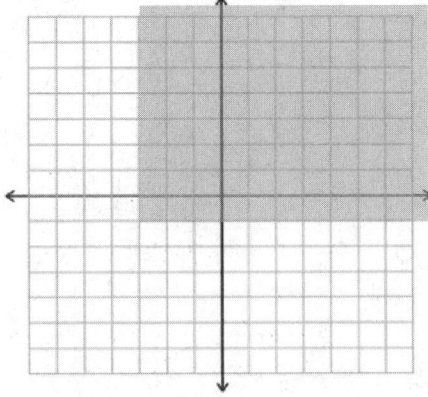

2. Region T is shown to the right.

    a. Write the coordinates of the vertices.

    b. Write an inequality that describes the region.

    c. What is the length of the diagonals?

    d. Give the coordinates of a point that is both in the region and on one of the diagonals.

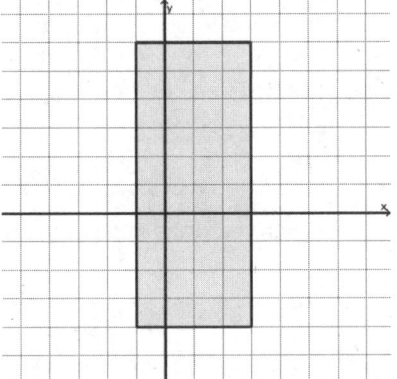

3. Jack wants to plant a garden in his backyard. His yard is 120 feet wide and 80 feet deep. He wants to plant a garden that is 20 feet by 30 feet.

    a. Set up a grid for the backyard, and place the garden on the grid. Explain why you placed your garden in its place on the grid.

    b. Write a system of inequalities to describe the garden.

    c. Write the equation of three lines that would go through the region that he could plant on, and explain your choices.

EUREKA
MATH

4. Given the trapezoidal region shown to the right:

    a. Write the system of inequalities describing the region.

    b. Translate the region to the right 3 units and down 2 units. Write the system of inequalities describing the translated region.

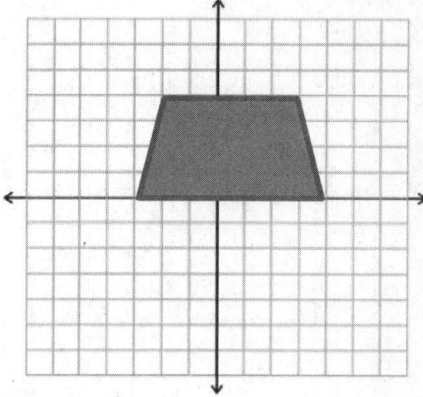

Challenge Problems:

5. Given the triangular region shown with vertices $A(-2,-1)$, $B(4,5)$, and $C(5,-1)$:

    a. Describe the systems of inequalities that describe the region enclosed by the triangle.

    b. Rotate the region $90°$ counterclockwise about Point A. How will this change the coordinates of the vertices?

    c. Write the system of inequalities that describe the region enclosed in the rotated triangle.

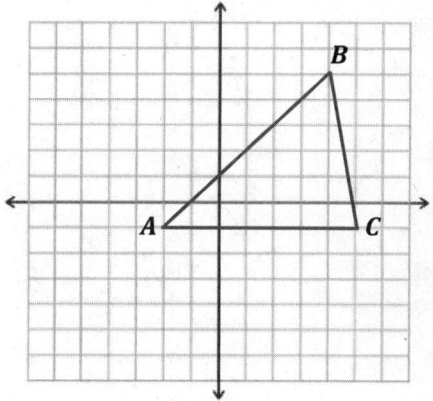

6. Write a system of inequalities for the region shown.

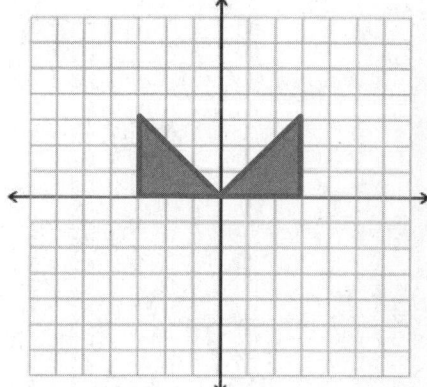

© 2015 Great Minds. eureka-math.org
GEO-M3-SE-B2-1.3.0-10.2015

EUREKA
MATH™

This page intentionally left blank

Lesson 3: Lines That Pass Through Regions

Classwork

Opening Exercise

How can we use the Pythagorean theorem to find the length of $\overline{AB}$, or in other words, the distance between $A(-2,1)$ and $B(3,3)$? Find the distance between A and B.

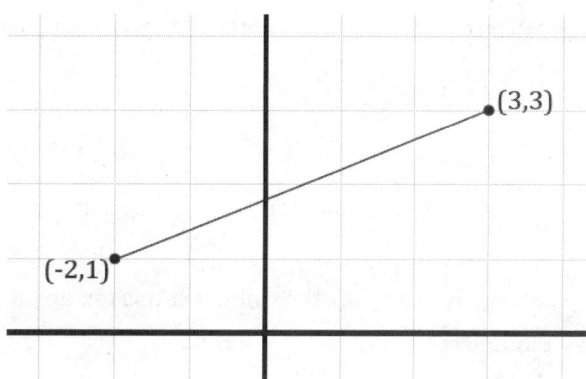

Example 1

Consider the rectangular region:

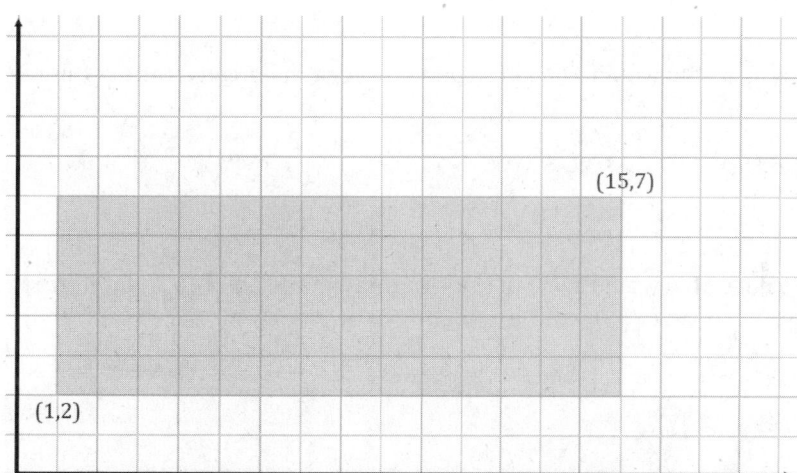

a.   Does a line of slope 2 passing through the origin intersect this rectangular region? If so, which boundary points of the rectangle does it intersect? Explain how you know.

b.   Does a line of slope $\frac{1}{2}$ passing through the origin intersect this rectangular region? If so, which boundary points of the rectangle does it intersect?

c.   Does a line of slope $\frac{1}{3}$ passing through the origin intersect this rectangular region? If so, which boundary points of the rectangle does it intersect?

d.   A line passes through the origin and the lower right vertex of the rectangle. Does the line pass through the interior of the rectangular region or the boundary of the rectangular region? Does the line pass through both?

e.   For which values of m would a line of slope m through the origin intersect this region?

f.   For which values of m would a line of slope m through the point $(0,1)$ intersect this region?

Example 2

Consider the triangular region in the plane given by the triangle with vertices $A(0,0)$, $B(2,6)$, and $C(4,2)$.

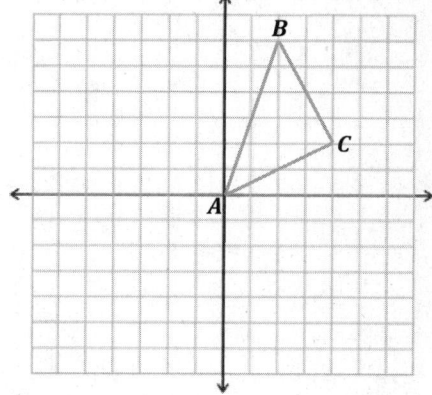

a. The horizontal line $y = 2$ intersects this region. What are the coordinates of the two boundary points it intersects? What is the length of the horizontal segment within the region along this line?

b. Graph the line $3x - 2y = 5$. Find the points of intersection with the boundary of the triangular region, and label them as X and Y.

c. What is the length of the $\overline{XY}$?

d. A robot starts at position $(1,3)$ and moves vertically downward toward the x-axis at a constant speed of 0.2 units per second. When will it hit the lower boundary of the triangular region that falls in its vertical path?

EUREKA
MATH™

Lesson 3:   Lines That Pass Through Regions

S.15

© 2015 Great Minds. eureka-math.org
GEO-M3-SE-B2-1.3.0-10.2015

Exercise

Consider the given rectangular region:

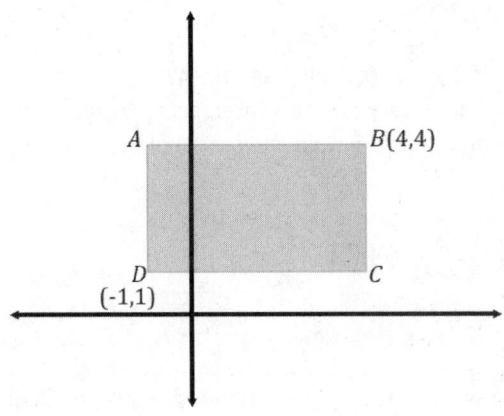

a.   Draw lines that pass through the origin and through each of the vertices of the rectangular region. Do each of the four lines cross multiple points in the region? Explain.

b.   Write the equation of a line that does not intersect the rectangular region at all.

c.   A robot is positioned at D and begins to move in a straight line with slope $m = 1$. When it intersects with a boundary, it then reorients itself and begins to move in a straight line with a slope of $m = -\frac{1}{2}$. What is the location of the next intersection the robot makes with the boundary of the rectangular region?

d.   What is the approximate distance of the robot's path in part (c)?

EUREKA
MATH™

Problem Set

1. A line intersects a triangle at least once, but not at any of its vertices. What is the maximum number of sides that a line can intersect a triangle? Similarly, a square? A convex quadrilateral? A quadrilateral, in general?

2. Consider the rectangular region:

   a. What boundary points does a line through the origin with a slope of -2 intersect? What is the length of the segment within this region along this line?

   b. What boundary points does a line through the origin with a slope of 3 intersect? What is the length of the segment within this region along this line?

   c. What boundary points does a line through the origin with a slope of $-\frac{1}{5}$ intersect?

   d. What boundary points does a line through the origin with a slope of $\frac{1}{4}$ intersect?

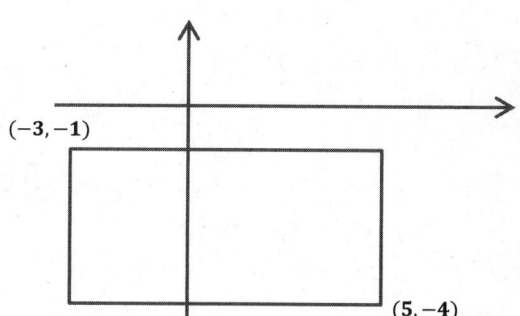

3. Consider the triangular region in the plane given by the triangle $(-1,3)$, $(1,-2)$, and $(-3,-3)$.

   a. The horizontal line $y = 1$ intersects this region. What are the coordinates of the two boundary points it intersects? What is the length of the horizontal segment within the region along this line?

   b. What is the length of the section of the line $2x + 3y = -4$ that lies within this region?

   c. If a robot starts at $(-1,3)$ and moves vertically downward at a constant speed of 0.75 units per second, when will it hit the lower boundary of the triangular region?

   d. If the robot starts at $(1,-2)$ and moves horizontally left at a constant speed of 0.6 units per second, when will it hit the left boundary of the triangular region?

4. A computer software exists so that the cursor of the program begins and ends at the origin of the plane. A program is written to draw a triangle with vertices $A\ (1,4)$, $B(6,2)$, and $C(3,1)$ so that the cursor only moves in straight lines and travels from the origin to A, then to B, then to C, then to A, and then back "home" to the origin.

   a. Sketch the cursor's path (i.e., sketch the entire path from when it begins until when it returns "home").

   b. What is the approximate total distance traveled by the cursor?

   c. Assume the cursor is positioned at B and is moving horizontally toward the y-axis at $\frac{2}{3}$ units per second. How long will it take to reach the boundary of the triangle?

5. An equilateral triangle with side length 1 is placed in the first quadrant so that one of its vertices is at the origin, and another vertex is on the x-axis. A line passes through the point half the distance between the endpoints on one side and half the distance between the endpoints on the other side.

   a. Draw a picture that satisfies these conditions.

   b. Find the equation of the line that you drew.

This page intentionally left blank

Lesson 4: Designing a Search Robot to Find a Beacon

Classwork

Opening Exercise

Write the equation of the line that satisfies the following conditions:

   a.   Has a slope of $m = -\frac{1}{4}$ and passes through the point $(0, -5)$.

   b. Passes through the points $(1, 3)$ and $(-2, -1)$.

Exploratory Challenge

A search robot is sweeping through a flat plane in search of the homing beacon that is emitting a signal. (A homing beacon is a tracking device that sends out signals to identify the location). Programmers have set up a coordinate system so that their location is the origin, the positive x-axis is in the direction of east, and the positive y-axis is in the direction of north. The robot is currently 600 units south of the programmers' location and is moving in an approximate northeast direction along the line $y = 3x - 600$.

Along this line, the robot hears the loudest "ping" at the point $(400, 600)$. It detects this ping coming from approximately a southeast direction. The programmers have the robot return to the point $(400, 600)$. What is the equation of the path the robot should take from here to reach the beacon?

Begin by sketching the location of the programmers and the path traveled by the robot on graph paper; then, shade the general direction the ping is coming from.

Notes:

Example

The line segment connecting (3,7) to (10,1) is rotated clockwise 90° about the point (3,7).

    a.   Plot the segment.

    b.   Where will the rotated endpoint land?

    c.   Now rotate the original segment 90° counterclockwise. Before using a sketch, predict the coordinates of the rotated endpoint using what you know about the perpendicular slope of the rotated segment.

Exercise

The point (a, b) is labeled below:

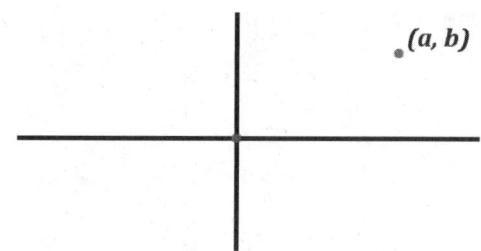

a. Using a and b, describe the location of (a, b) after a 90° counterclockwise rotation about the origin. Draw a rough sketch to justify your answer.

b. If the rotation was *clockwise* about the origin, what is the rotated location of (a, b) in terms of a and b? Draw a rough sketch to justify your answer.

c. What is the slope of the line through the origin and (a, b)? What is the slope of the perpendicular line through the origin?

d. What do you notice about the relationship between the slope of the line through the origin and (a, b) and the slope of the perpendicular line?

Lesson 4:   Designing a Search Robot to Find a Beacon

S.21

Problem Set

1. Find the new coordinates of point $(0,4)$ if it rotates:

   a. 90° counterclockwise.

   b. 90° clockwise.

   c. 180° counterclockwise.

   d. 270° clockwise.

2. What are the new coordinates of the point $(-3, -4)$ if it is rotated about the origin:

   a. Counterclockwise 90°?

   b. Clockwise 90°?

3. Line segment ST connects points $S(7,1)$ and $T(2,4)$.

   a. Where does point T land if the segment is rotated 90° counterclockwise about S?

   b. Where does point T land if the segment is rotated 90° clockwise about S?

   c. What is the slope of the original segment?

   d. What is the slope of the rotated segments?

4. Line segment VW connects points $V(1,0)$ and $W(5, -3)$.

   a. Where does point W land if the segment is rotated 90° counterclockwise about V?

   b. Where does point W land if the segment is rotated 90° clockwise about V?

   c. Where does point V land if the segment is rotated 90° counterclockwise about W?

   d. Where does point V land if the segment is rotated 90° clockwise about W?

5. If the slope of a line is 0, what is the slope of a line perpendicular to it? If the line has slope 1, what is the slope of a line perpendicular to it?

6. If a line through the origin has a slope of 2, what is the slope of the line through the origin that is perpendicular to it?

7. A line through the origin has a slope of $\frac{1}{3}$. Carlos thinks the slope of a perpendicular line at the origin will be 3. Do you agree? Explain why or why not.

8. Could a line through the origin perpendicular to a line through the origin with slope $\frac{1}{2}$ pass through the point $(-1,4)$? Explain how you know.

Lesson 5: Criterion for Perpendicularity

Classwork

Opening Exercise

In right triangle ABC, find the missing side.

    a.   If $AC = 9$ and $CB = 12$, what is AB? Explain how you know.

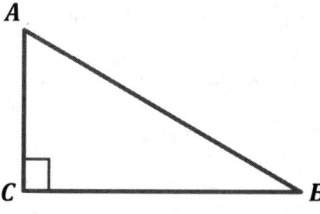

    b.   If $AC = 5$ and $AB = 13$, what is CB?

    c.   If $AC = CB$ and $AB = 2$, what is AC (and CB)?

Exercise 1

1.   Use the grid on the right.

    a.   Plot points $O(0,0)$, $P(3,-1)$, and $Q(2,3)$ on the coordinate plane.

    b.   Determine whether $\overline{OP}$ and $\overline{OQ}$ are perpendicular. Support your findings.

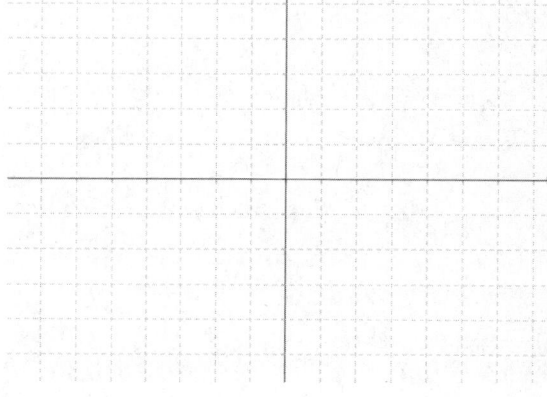

Example 2

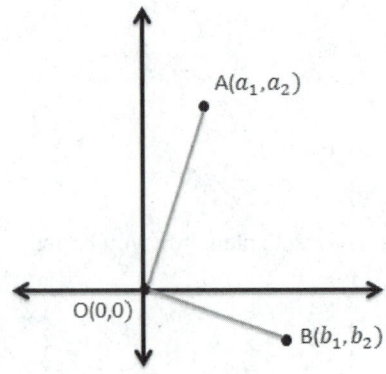

Exercises 2–3

2. Given points $O(0,0)$, $A(6,4)$, $B(24,-6)$, $C(1,4)$, $P(2,-3)$, $S(-18,-12)$, $T(-3,-12)$, $U(-8,2)$, and $W(-6,9)$, find all pairs of segments from the list below that are perpendicular. Support your answer.

$\overline{OA}$, $\overline{OB}$, $\overline{OC}$, $\overline{OP}$, $\overline{OS}$, $\overline{OT}$, $\overline{OU}$, and $\overline{OW}$

3. The points $O(0,0)$, $A(-4,1)$, $B(-3,5)$, and $C(1,4)$ are the vertices of parallelogram $OABC$. Is this parallelogram a rectangle? Support your answer.

Problem Set

1. Prove using the Pythagorean theorem that $\overline{AC}$ is perpendicular to $\overline{AB}$ given points $A(-2,-2)$, $B(5,-2)$, and $C(-2,22)$.

2. Using the general formula for perpendicularity of segments through the origin and $(90,0)$, determine if $\overline{OA}$ and $\overline{OB}$ are perpendicular.

    a. $A(-3,-4)$, $B(4,3)$

    b. $A(8,9)$, $B(18,-16)$

3. Given points $O(0,0)$, $S(2,7)$, and $T(7,-2)$, where $\overline{OS}$ is perpendicular to $\overline{OT}$, will the images of the segments be perpendicular if the three points O, S, and T are translated four units to the right and eight units up? Explain your answer.

4. In Example 1, we saw that $\overline{OA}$ was perpendicular to $\overline{OB}$ for $O(0,0)$, $A(6,4)$, and $B(-2,3)$. Suppose we are now given the points $P(5,5)$, $Q(11,9)$, and $R(3,8)$. Are segments $\overline{PQ}$ and $\overline{PR}$ perpendicular? Explain without using triangles or the Pythagorean theorem.

5. Challenge: Using what we learned in Exercise 2, given points $C(c_1, c_2)$, $A(a_1, a_2)$, and $B(b_1, b_2)$, what is the general condition of a_1, a_2, b_1, b_2, c_1, and c_2 that ensures $\overline{CA}$ and $\overline{CB}$ are perpendicular?

6. A robot that picks up tennis balls is on a straight path from $(8,6)$ toward a ball at $(-10,-5)$. The robot picks up a ball at $(-10,-5)$ and then turns $90°$ right. What are the coordinates of a point that the robot can move toward to pick up the last ball?

7. Gerry thinks that the points $(4,2)$ and $(-1,4)$ form a line perpendicular to a line with slope 4. Do you agree? Why or why not?

Lesson 6: Segments That Meet at Right Angles

Classwork

Opening Exercise

Carlos thinks that the segment having endpoints $A(0,0)$ and $B(6,0)$ is perpendicular to the segment with endpoints $A(0,0)$ and $C(-2,0)$. Do you agree? Why or why not?

Working with a partner, given $A(0,0)$ and $B(3,-2)$, find the coordinates of a point C so that $\overline{AC} \perp \overline{AB}$.

Example

Given points $A(2,2)$, $B(10,16)$, $C(-3,1)$, and $D(4,-3)$, are $\overline{AB}$ and $\overline{CD}$ perpendicular? Are the lines containing the segments perpendicular? Explain.

Exercises

1. Given $A(a_1, a_2)$, $B(b_1, b_2)$, $C(c_1, c_2)$, and $D(d_1, d_2)$, find a general formula in terms of a_1, a_2, b_1, b_2, c_1, c_2, d_1, and d_2 that will let us determine whether $\overline{AB}$ and $\overline{CD}$ are perpendicular.

2. Recall the Opening Exercise of Lesson 4 in which a robot is traveling along a linear path given by the equation $y = 3x - 600$. The robot hears a ping from a homing beacon when it reaches the point $F(400, 600)$ and turns to travel along a linear path given by the equation $y - 600 = -\frac{1}{3}(x - 400)$. If the homing beacon lies on the x-axis, what is its exact location? (Use your own graph paper to visualize the scenario.)

    a. If point E is the y-intercept of the original equation, what are the coordinates of point E?

    b. What are the endpoints of the original segment of motion?

    c. If the beacon lies on the x-axis, what is the y-value of this point, G?

    d. Translate point F to the origin. What are the coordinates of E', F', and G'?

    e. Use the formula derived in this lesson to determine the coordinates of point G.

Lesson 6:   Segments That Meet at Right Angles

EUREKA
MATH™

© 2015 Great Minds. eureka-math.org
GEO-M3-SE-B2-1.3.0-10.2015

3. A triangle in the coordinate plane has vertices $A(0,10)$, $B(-8,8)$, and $C(-3,5)$. Is it a right triangle? If so, at which vertex is the right angle? (Hint: Plot the points, and draw the triangle on a coordinate plane to help you determine which vertex is the best candidate for the right angle.)

4. $A(-7,1)$, $B(-1,3)$, $C(5,-5)$, and $D(-5,-5)$ are vertices of a quadrilateral. If $\overline{AC}$ bisects $\overline{BD}$, but $\overline{BD}$ does not bisect $\overline{AC}$, determine whether $ABCD$ is a kite.

Problem Set

1. Are the segments through the origin and the points listed perpendicular? Explain.

   a. $A(9,10)$, $B(10,9)$

   b. $C(9,6)$, $D(4,-6)$

2. Given $M(5,2)$, $N(1,-4)$, and L listed below, are $\overline{LM}$ and $\overline{MN}$ perpendicular? Translate M to the origin, write the coordinates of the images of the points, and then explain without using slope.

   a. $L(-1,6)$

   b. $L(11,-2)$

   c. $L(9,8)$

3. Is triangle PQR, where $P(-7,3)$, $Q(-4,7)$, and $R(1,-3)$, a right triangle? If so, which angle is the right angle? Justify your answer.

4. A quadrilateral has vertices $(2 + \sqrt{2}, -1)$, $(8 + \sqrt{2}, 3)$, $(6 + \sqrt{2}, 6)$, and $(\sqrt{2}, 2)$. Prove that the quadrilateral is a rectangle.

5. Given points $G(-4,1)$, $H(3,2)$, and $I(-2,-3)$, find the x-coordinate of point J with y-coordinate 4 so that the $\overleftrightarrow{GH}$ and $\overrightarrow{IJ}$ are perpendicular.

6. A robot begins at position $(-80,45)$ and moves on a path to $(100,-60)$. It turns $90°$ counterclockwise.

   a. What point with y-coordinate 120 is on this path?

   b. Write an equation of the line after the turn.

   c. If it stops to charge on the x-axis, what is the location of the charger?

7. Determine the missing vertex of a right triangle with vertices $(6,2)$ and $(5,5)$ if the third vertex is on the y-axis. Verify your answer by graphing.

8. Determine the missing vertex for a rectangle with vertices $(3,-2)$, $(5,2)$, and $(-1,5)$, and verify by graphing. Then, answer the questions that follow.

   a. What is the length of the diagonal?

   b. What is a point on both diagonals in the interior of the figure?

9. Leg $\overline{AB}$ of right triangle ABC has endpoints $A(1,3)$ and $B(6,-1)$. Point $C(x,y)$ is located in Quadrant IV.

   a. Use the perpendicularity criterion to determine at which vertex the right angle is located. Explain your reasoning.

   b. Determine the range of values that x is limited to and why.

   c. Find the coordinates of point C if they are both integers.

Lesson 7: Equations for Lines Using Normal Segments

Classwork

Opening Exercise

The equations given are in standard form. Put each equation in slope-intercept form. State the slope and the y-intercept.

1. $6x + 3y = 12$

2. $5x + 7y = 14$

3. $2x - 5y = -7$

Example

Given $A(5, -7)$ and $B(8,2)$:

   a.   Find an equation for the line through A and perpendicular to $\overline{AB}$.

   b.   Find an equation for the line through B and perpendicular to $\overline{AB}$.

Exercises

1. Given $U(-4, -1)$ and $V(7,1)$:

    a. Write an equation for the line through U and perpendicular to $\overline{UV}$.

    b. Write an equation for the line through V and perpendicular to $\overline{UV}$.

2. Given $S(5, -4)$ and $T(-8,12)$:

    a. Write an equation for the line through S and perpendicular to $\overline{ST}$.

    b. Write an equation for the line through T and perpendicular to $\overline{ST}$.

EUREKA
MATH™

Closing

Describe the characteristics of a normal segment.

Every equation of a line through a given point (a, b) has the form $A(x - a) + B(y - b) = 0$. Explain how the values of A and B are obtained.

Problem Set

1. Given points $C(-4,3)$ and $D(3,3)$:

    a. Write the equation of the line through C and perpendicular to $\overline{CD}$.

    b. Write the equation of the line through D and perpendicular to $\overline{CD}$.

2. Given points $N(7,6)$ and $M(7,-2)$:

    a. Write the equation of the line through M and perpendicular to $\overline{MN}$.

    b. Write the equation of the line through N and perpendicular to $\overline{MN}$.

3. The equation of a line is given by the equation $8(x-4) + 3(y+2) = 0$.

    a. What are the coordinates of the image of the endpoint of the normal segment that does not lie on the line? Explain your answer.

    b. What translation occurred to move the point of perpendicularity to the origin?

    c. What were the coordinates of the original point of perpendicularity? Explain your answer.

    d. What were the endpoints of the original normal segment?

4. A coach is laying out lanes for a race. The lanes are perpendicular to a segment of the track such that one endpoint of the segment is $(2,50)$, and the other is $(20,65)$. What are the equations of the lines through the endpoints?

Lesson 7:   Equations for Lines Using Normal Segments

EUREKA
MATH™

© 2015 Great Minds. eureka-math.org
GEO-M3-SE-B2-1.3.0-10.2015

Lesson 8: Parallel and Perpendicular Lines

Classwork

Exercise 1

1.

    a. Write an equation of the line that passes through the origin that intersects the line $2x + 5y = 7$ to form a right angle.

    b. Determine whether the lines given by the equations $2x + 3y = 6$ and $y = \frac{3}{2}x + 4$ are perpendicular. Support your answer.

    c. Two lines having the same y-intercept are perpendicular. If the equation of one of these lines is $y = -\frac{4}{5}x + 6$, what is the equation of the second line?

Example 2

    a. What is the relationship between two coplanar lines that are perpendicular to the same line?

b. Given two lines, l_1 and l_2, with equal slopes and a line k that is perpendicular to one of these two parallel lines, l_1:

    i.   What is the relationship between line k and the other line, l_2?

    ii.   What is the relationship between l_1 and l_2?

Exercises 2–7

2. Given a point $(-3,6)$ and a line $y = 2x - 8$:

    a.   What is the slope of the line?

    b.   What is the slope of any line parallel to the given line?

    c.   Write an equation of a line through the point and parallel to the line.

    d.   What is the slope of any line perpendicular to the given line? Explain.

EUREKA
MATH™

3. Find an equation of a line through $(0, -7)$ and parallel to the line $y = \frac{1}{2}x + 5$.

    a. What is the slope of any line parallel to the given line? Explain your answer.

    b. Write an equation of a line through the point and parallel to the line.

    c. If a line is perpendicular to $y = \frac{1}{2}x + 5$, will it be perpendicular to $x - 2y = 14$? Explain.

4. Find an equation of a line through $\left(\sqrt{3}, \frac{1}{2}\right)$ parallel to the line:

    a. $x = -9$

    b. $y = -\sqrt{7}$

    c. What can you conclude about your answer in parts (a) and (b)?

5. Find an equation of a line through $\left(-\sqrt{2}, \pi\right)$ parallel to the line $x - 7y = \sqrt{5}$.

6. Recall that our search robot is moving along the line $y = 3x - 600$ and wishes to make a right turn at the point $(400,600)$. Find an equation for the perpendicular line on which the robot is to move. Verify that your line intersects the x-axis at $(2200,0)$.

7. A robot, always moving at a constant speed of 2 units per second, starts at position $(20,50)$ on the coordinate plane and heads in a southeast direction along the line $3x + 4y = 260$. After 15 seconds, it turns clockwise $90°$ and travels in a straight line in this new direction.

    a. What are the coordinates of the point at which the robot made the turn? What might be a relatively straightforward way of determining this point?

    b. Find an equation for the second line on which the robot traveled.

    c. If, after turning, the robot travels for 20 seconds along this line and then stops, how far will it be from its starting position?

    d. What is the equation of the line the robot needs to travel along in order to now return to its starting position? How long will it take for the robot to get there?

EUREKA
MATH™

© 2015 Great Minds. eureka-math.org
GEO-M3-SE-B2-1.3.0-10.2015

Problem Set

1. Write the equation of the line through $(-5, 3)$ and:

    a. Parallel to $x = -1$.

    b. Perpendicular to $x = -1$.

    c. Parallel to $y = \frac{3}{5}x + 2$.

    d. Perpendicular to $y = \frac{3}{5}x + 2$.

2. Write the equation of the line through $\left(\sqrt{3}, \frac{5}{4}\right)$ and:

    a. Parallel to $y = 7$.

    b. Perpendicular to $y = 7$.

    c. Parallel to $\frac{1}{2}x - \frac{3}{4}y = 10$.

    d. Perpendicular to $\frac{1}{2}x - \frac{3}{4}y = 10$.

3. A vacuum robot is in a room and charging at position $(0, 5)$. Once charged, it begins moving on a northeast path at a constant speed of $\frac{1}{2}$ foot per second along the line $4x - 3y = -15$. After 60 seconds, it turns right $90°$ and travels in the new direction.

    a. What are the coordinates of the point at which the robot made the turn?

    b. Find an equation for the second line on which the robot traveled.

    c. If after turning, the robot travels 80 seconds along this line, what is the distance between the starting position and the robot's current position?

    d. What is the equation of the line the robot needs to travel along in order to return and recharge? How long will it take the robot to get there?

4. Given the statement $\overleftrightarrow{AB}$ is parallel to $\overleftrightarrow{DE}$, construct an argument for or against this statement using the two triangles shown.

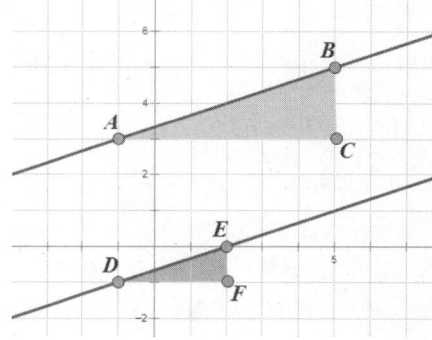

EUREKA
MATH™

5. Recall the proof we did in Example 1: Let l_1 and l_2 be two non-vertical lines in the Cartesian plane. l_1 and l_2 are perpendicular if and only if their slopes are negative reciprocals of each other. In class, we looked at the case where both y-intercepts were not zero. In Lesson 5, we looked at the case where both y-intercepts were equal to zero, when the vertex of the right angle was at the origin. Reconstruct the proof for the case where one line has a y-intercept of zero, and the other line has a nonzero y-intercept.

6. Challenge: Reconstruct the proof we did in Example 1 if one line has a slope of zero.

© 2015 Great Minds. eureka-math.org
GEO-M3-SE-B2-1.3.0-10.2015

Lesson 9: Perimeter and Area of Triangles in the Cartesian Plane

Classwork

Opening Exercise

Find the area of the shaded region.

a.

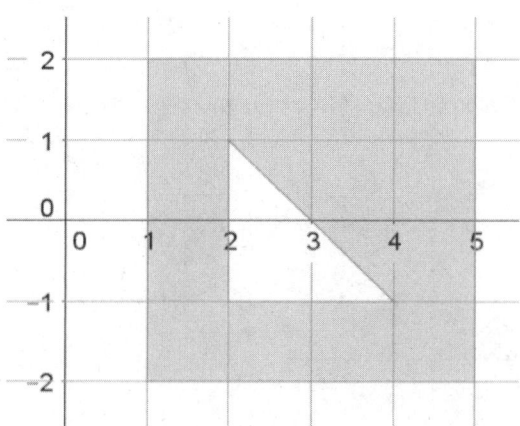

b.

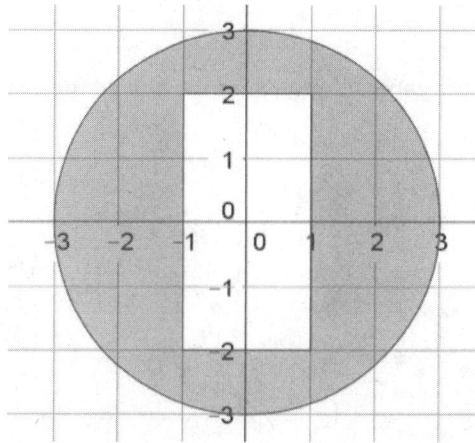

Example

Consider a triangular region in the plane with vertices $O(0,0)$, $A(5,2)$, and $B(3,4)$. What is the perimeter of the triangular region?

What is the area of the triangular region?

Find the general formula for the area of the triangle with vertices $O(0,0)$, $A(x_1, y_1)$, and $B(x_2, y_2)$, as shown.

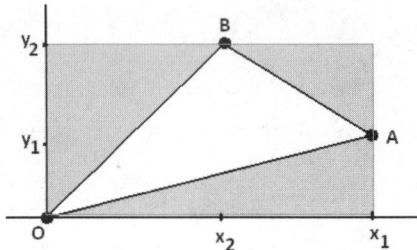

Does the formula work for this triangle?

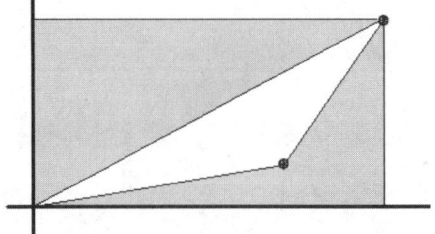

EUREKA
MATH™

Exercise

Find the area of the triangles with vertices listed, first by finding the area of the rectangle enclosing the triangle and subtracting the area of the surrounding triangles, then by using the formula $\frac{1}{2}(x_1 y_2 - x_2 y_1)$.

a. $O(0,0), A(5,6), B(4,1)$

b. $O(0,0), A(3,2), B(-2,6)$

c. $O(0,0), A(5,-3), B(-2,6)$

Problem Set

1. Use coordinates to compute the perimeter and area of each polygon.

   a.

   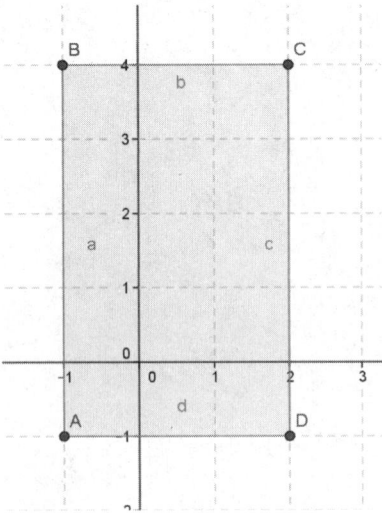

   b.

   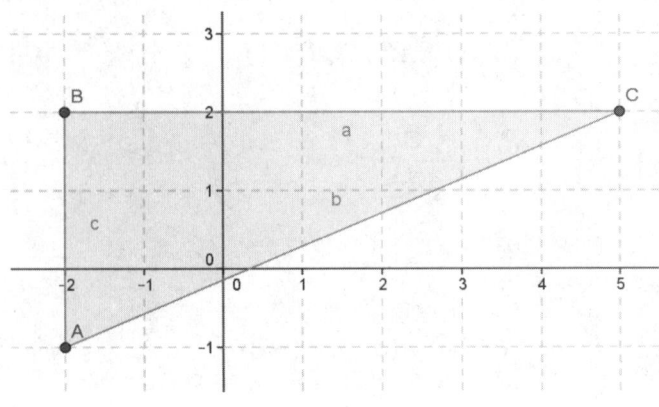

2. Given the figures below, find the area by decomposing into rectangles and triangles.

   a.

   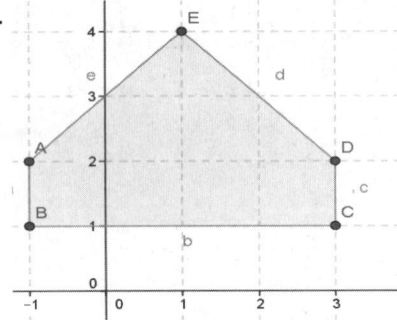

   b.

   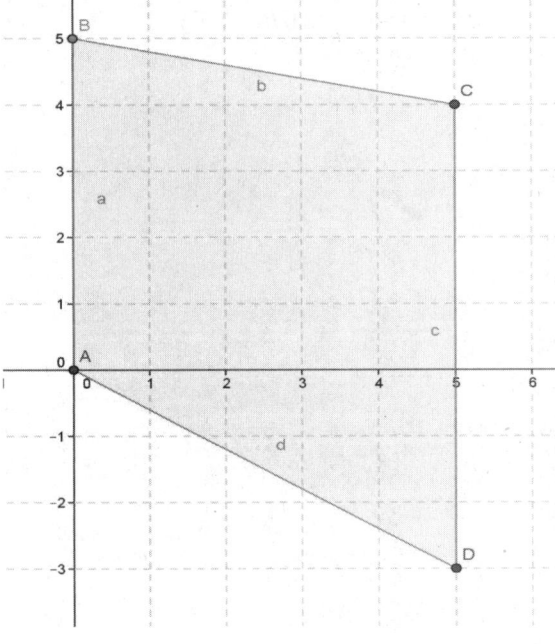

EUREKA
MATH™

3. Challenge: Find the area by decomposing the given figure into triangles.

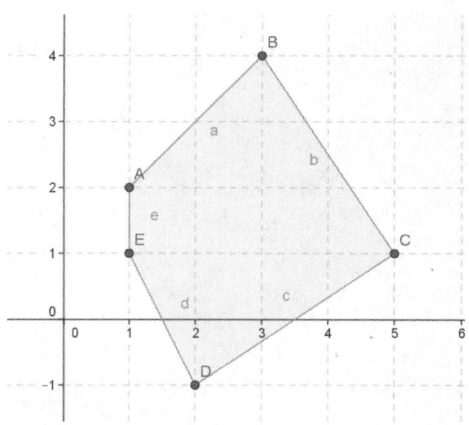

4. When using the shoelace formula to work out the area of △ ABC, we have some choices to make. For example, we can start at any one of the three vertices A, B, or C, and we can move either in a clockwise or counterclockwise direction. This gives six options for evaluating the formula.

Show that the shoelace formula obtained is identical for the three options that move in a clockwise direction (A to C to B or C to B to A or B to A to C) and identical for the three options in the reverse direction. Verify that the two distinct formulas obtained differ only by a minus sign.

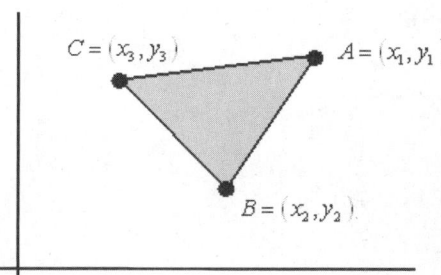

5. Suppose two triangles share a common edge. By translating and rotating the triangles, we can assume that the common edge lies along the x-axis with one endpoint at the origin.

a. Show that if we evaluate the shoelace formula for each triangle, both calculated in the same clockwise direction, then the answers are both negative.

b. Show that if we evaluate them both in a counterclockwise direction, then both are positive.

c. Explain why evaluating one in one direction and the second in the opposite direction, the two values obtained are opposite in sign.

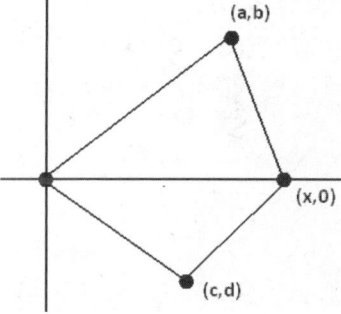

6. A textbook has a picture of a triangle with vertices $(3, 6)$ and $(5, 2)$. Something happened in printing the book, and the coordinates of the third vertex are listed as $(-1, \blacksquare)$. The answers in the back of the book give the area of the triangle as 6 square units.

a. What is the y-coordinate of the third vertex?

b. What if both coordinates were missing, but the area was known? Could you use algebra to find the third coordinate? Explain.

This page intentionally left blank

Lesson 10: Perimeter and Area of Polygonal Regions in the Cartesian Plane

Classwork

Opening Exercise

Find the area of the triangle given. Compare your answer and method to your neighbor's, and discuss differences.

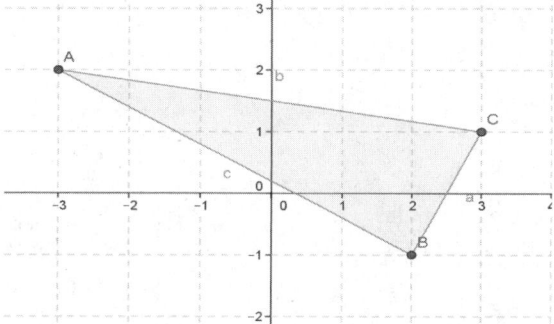

Exercises

1. Given rectangle $ABCD$:

    a. Identify the vertices.

    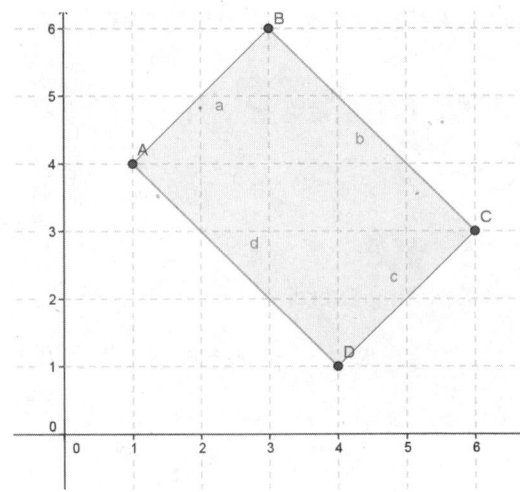

    b. Find the perimeter using the distance formula.

    c. Find the area using the area formula.

    d. List the vertices starting with A moving counterclockwise.

    e. Verify the area using the shoelace formula.

EUREKA
MATH™

2. Calculate the area and perimeter of the given quadrilateral using the shoelace formula.

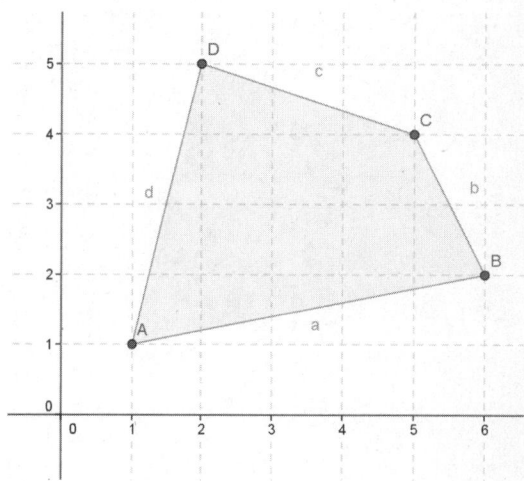

3. Break up the pentagon to find the area using Green's theorem. Compare your method with a partner.

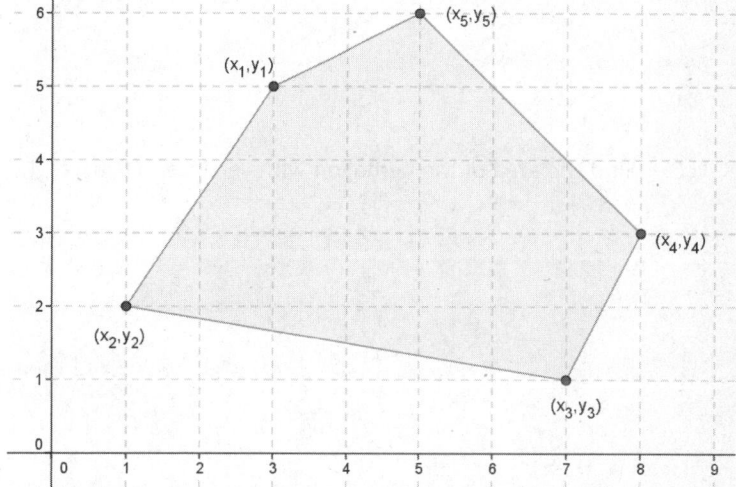

EUREKA
MATH™

4. Find the perimeter and the area of the quadrilateral with vertices $A(-3, 4)$, $B(4, 6)$, $C(2, -3)$, and $D(-4, -4)$.

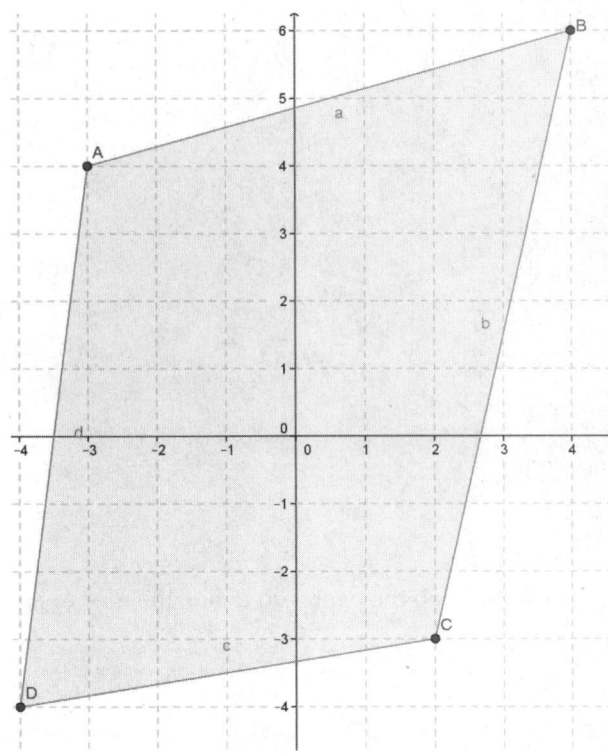

5. Find the area of the pentagon with vertices $A(5, 8)$, $B(4, -3)$, $C(-1, -2)$, $D(-2, 4)$, and $E(2, 6)$.

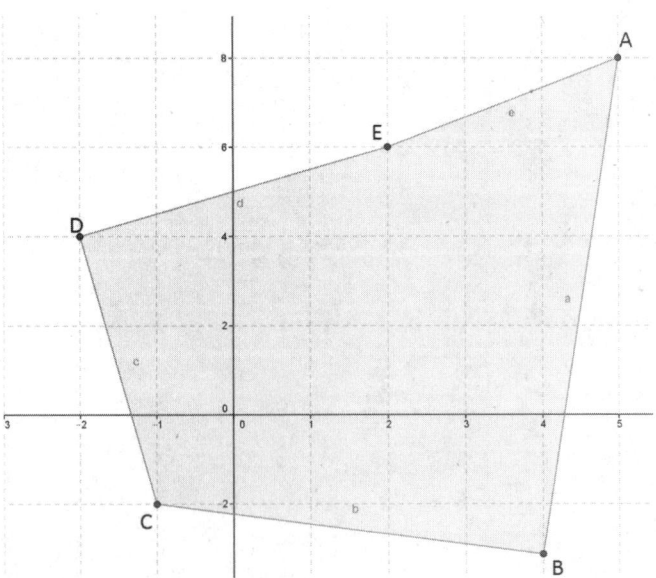

EUREKA
MATH™

© 2015 Great Minds. eureka-math.org
GEO-M3-SE-B2-1.3.0-10.2015

6. Find the area and perimeter of the hexagon shown.

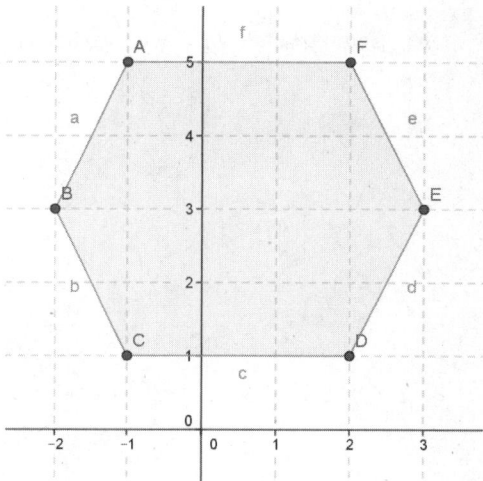

Problem Set

1. Given triangle ABC with vertices $(7, 4)$, $(1, 1)$, and $(9, 0)$:

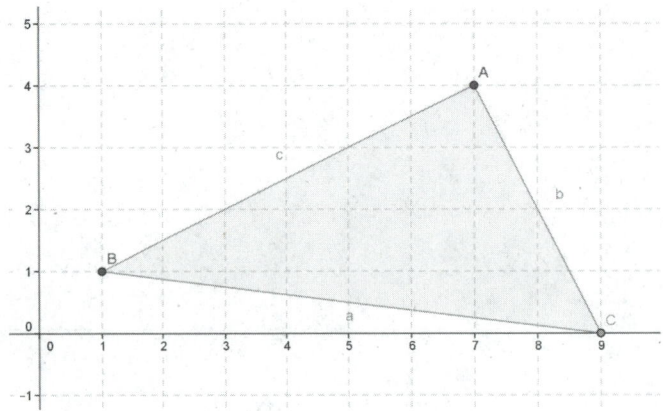

   a. Calculate the perimeter using the distance formula.

   b. Calculate the area using the traditional area formula.

   c. Calculate the area using the shoelace formula.

   d. Explain why the shoelace formula might be more useful and efficient if you were just asked to find the area.

2. Given triangle ABC and quadrilateral $DEFG$, describe how you would find the area of each and why you would choose that method, and then find the areas.

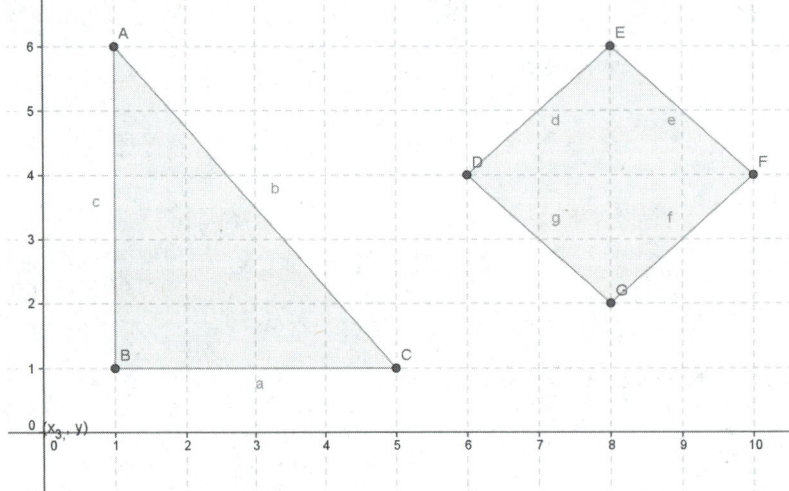

EUREKA
MATH™

3. Find the area and perimeter of quadrilateral $ABCD$ with vertices $A(6,5)$, $B(2,-4)$, $C(-5,2)$, and $D(-3,6)$.

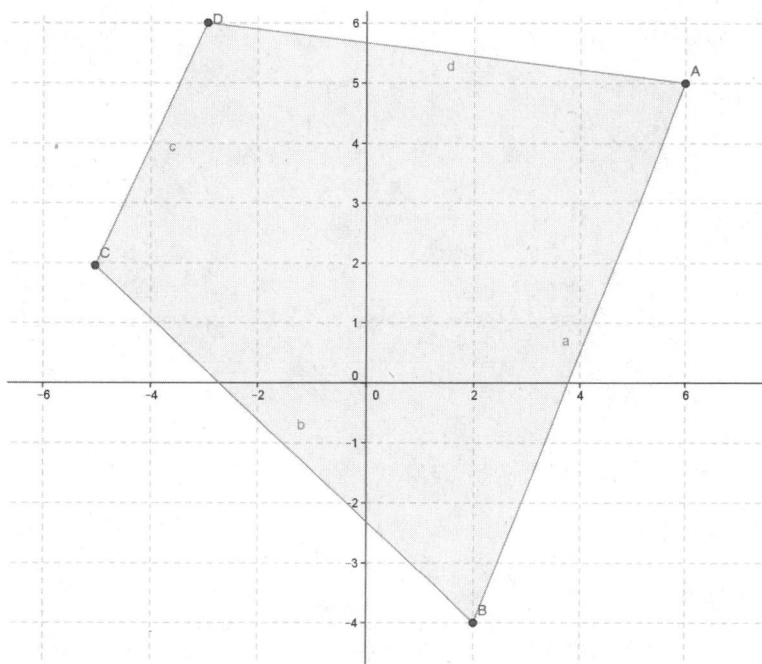

4. Find the area and perimeter of pentagon $ABCDE$ with vertices $A(2,6)$, $B(7,2)$, $C(3,-4)$, $D(-3,-2)$, and $E(-2,4)$.

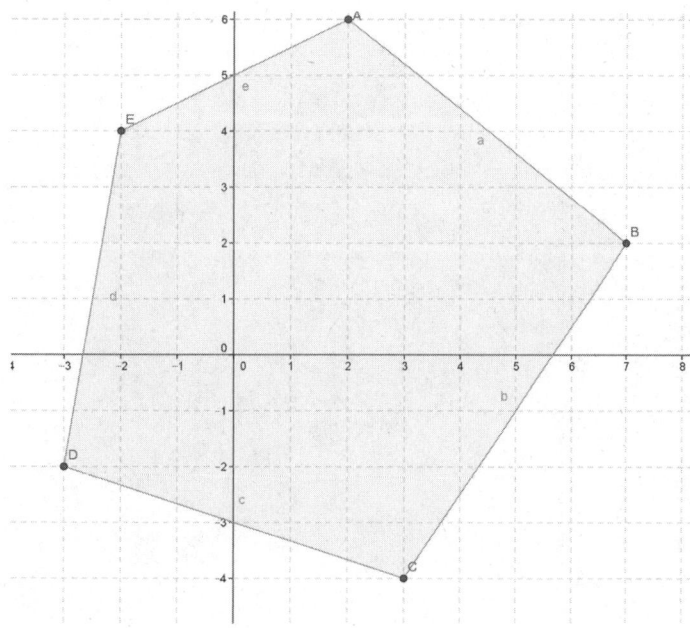

EUREKA
MATH™

5. Show that the shoelace formula (Green's theorem) used on the trapezoid shown confirms the traditional formula for the area of a trapezoid $\frac{1}{2}\left(b_1 + b_2\right) \cdot h$.

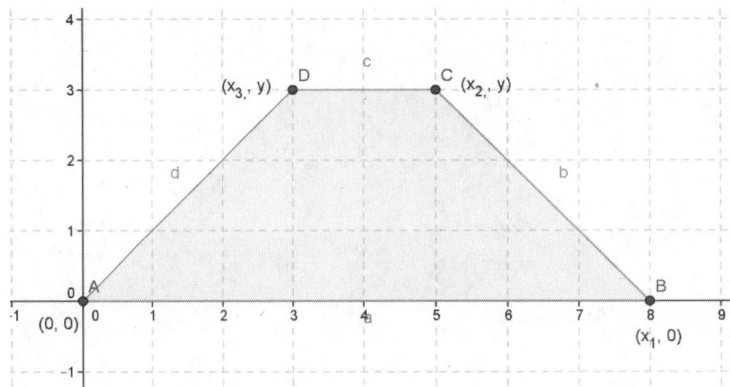

EUREKA
MATH™

Lesson 11: Perimeters and Areas of Polygonal Regions Defined by Systems of Inequalities

Classwork

Opening Exercise

Graph the following:

a. $y \leq 7$

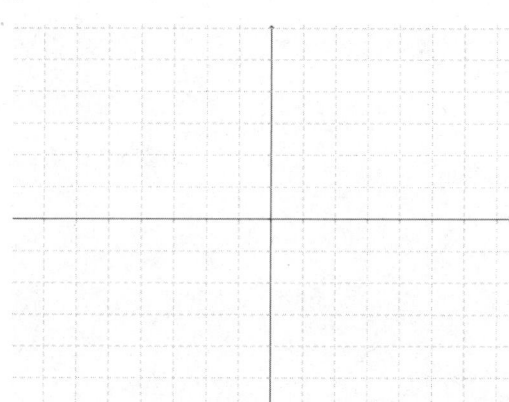

b. $x > -3$

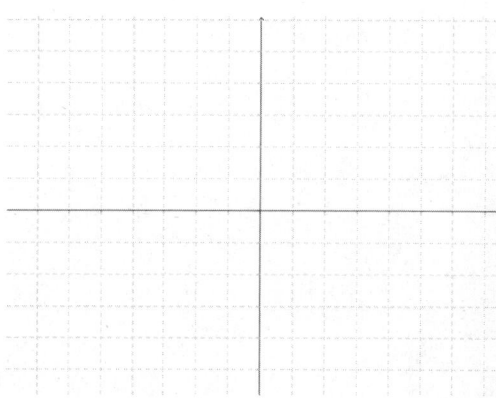

c. $y < \frac{1}{2}x - 4$

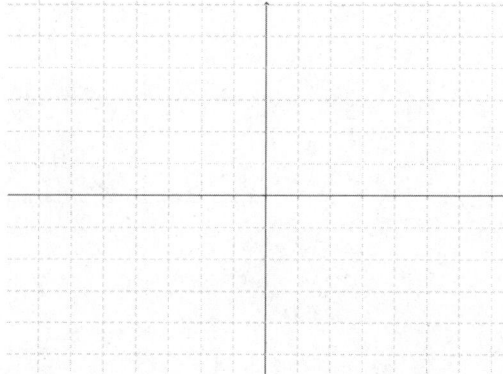

d. $y \geq -\frac{2}{3}x + 5$

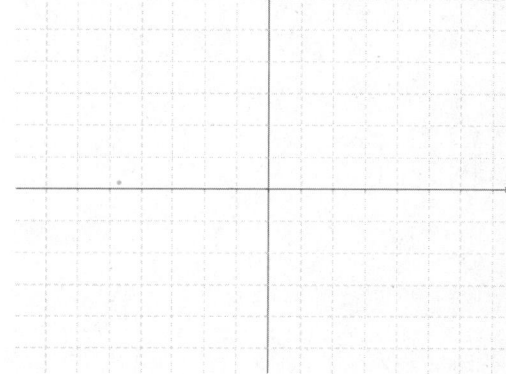

Example 1

A parallelogram with base of length b and height h can be situated in the coordinate plane, as shown. Verify that the shoelace formula gives the area of the parallelogram as bh.

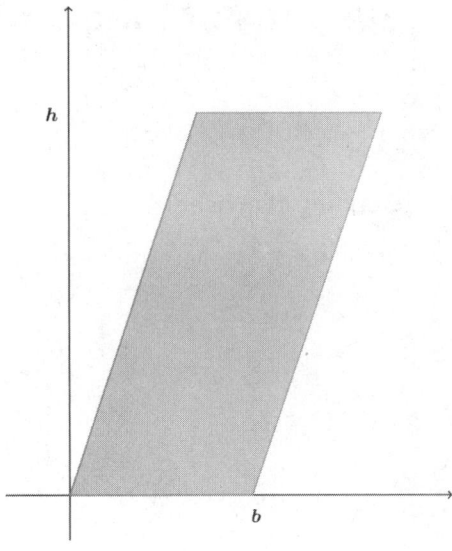

Example 2

A triangle with base b and height h can be situated in the coordinate plane, as shown. According to Green's theorem, what is the area of the triangle?

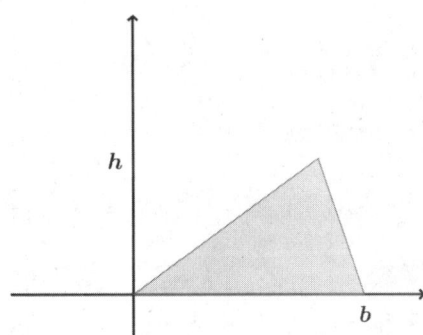

EUREKA
MATH™

Exercises

1. A quadrilateral region is defined by the system of inequalities below:

    $y \leq x + 6$ $y \leq -2x + 12$ $y \geq 2x - 4$ $y \geq -x + 2$

    a. Sketch the region.

    b. Determine the vertices of the quadrilateral.

    c. Find the perimeter of the quadrilateral region.

    d. Find the area of the quadrilateral region.

2. A quadrilateral region is defined by the system of inequalities below:

    $y \leq x + 5$ $y \geq x - 4$ $y \leq 4$ $y \geq -\dfrac{5}{4}x - 4$

    a. Sketch the region.

    b. Determine the vertices of the quadrilateral.

    c. Which quadrilateral is defined by these inequalities? How can you prove your conclusion?

    d. Find the perimeter of the quadrilateral region.

    e. Find the area of the quadrilateral region.

Problem Set

For Problems 1–2 below, identify the system of inequalities that defines the region shown.

1.

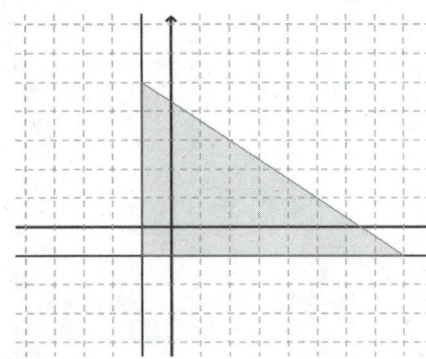

2.

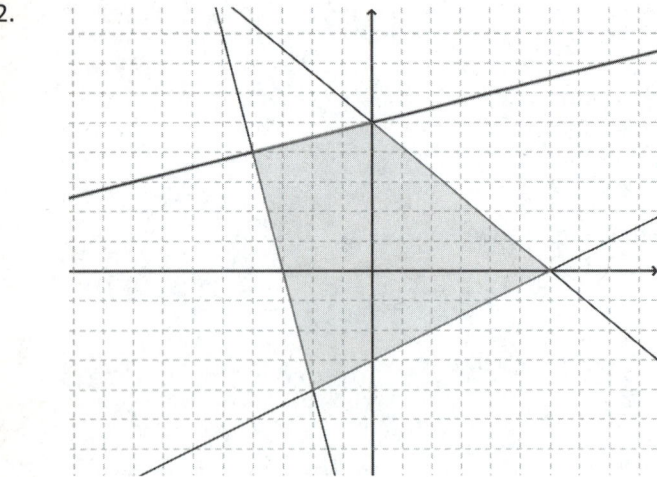

For Problems 3–5 below, a triangular or quadrilateral region is defined by the system of inequalities listed.

    a.   Sketch the region.

    b.   Determine the coordinates of the vertices.

    c.   Find the perimeter of the region rounded to the nearest hundredth if necessary.

    d.   Find the area of the region rounded to the nearest tenth if necessary.

3. $8x - 9y \geq -22$ $x + y \leq 10$ $5x - 12y \leq -1$

4. $x + 3y \geq 0$ $4x - 3y \geq 0$ $2x + y \leq 10$

5. $2x - 5y \geq -14$ $3x + 2y \leq 17$ $2x - y \leq 9$ $x + y \geq 0$

Lesson 12: Dividing Segments Proportionately

Classwork

Exercises

1. Find the midpoint of $\overline{ST}$ given $S(-2, 8)$ and $T(10, -4)$.

2. Find the point on the directed segment from $(-2, 0)$ to $(5, 8)$ that divides it in the ratio of $1:3$.

3. Given $\overline{PQ}$ and point R that lies on $\overline{PQ}$ such that point R lies $\dfrac{7}{9}$ of the length of $\overline{PQ}$ from point P along $\overline{PQ}$:

   a. Sketch the situation described.

   b. Is point R closer to P or closer to Q, and how do you know?

c. Use the given information to determine the following ratios:

i. $PR:PQ$

ii. $RQ:PQ$

iii. $PR:RQ$

iv. $RQ:PR$

d. If the coordinates of point P are $(0,0)$ and the coordinates of point R are $(14, 21)$, what are the coordinates of point Q?

4. A robot is at position $A(40, 50)$ and is heading toward the point $B(2000, 2000)$ along a straight line at a constant speed. The robot will reach point B in 10 hours.

a. What is the location of the robot at the end of the third hour?

b. What is the location of the robot five minutes before it reaches point B?

EUREKA
MATH™

c. If the robot keeps moving along the straight path at the same constant speed as it passes through point B, what will be its location at the twelfth hour?

d. Compare the value of the abscissa (x-coordinate) to the ordinate (y-coordinate) before, at, and after the robot passes point B.

e. Could you have predicted the relationship that you noticed in part (d) based on the coordinates of points A and B?

Problem Set

1. Given $F(0, 2)$ and $G(2, 6)$, if point S lies $\dfrac{5}{12}$ of the way along $\overline{FG}$, closer to F than to G, find the coordinates of S. Then verify that this point lies on $\overline{FG}$.

2. Point C lies $\dfrac{5}{6}$ of the way along $\overline{AB}$, closer to B than to A. If the coordinates of point A are $(12, 5)$ and the coordinates of point C are $(9.5, -2.5)$, what are the coordinates of point B?

3. Find the point on the directed segment from $(-3, -2)$ to $(4, 8)$ that divides it into a ratio of $3 : 2$.

4. A robot begins its journey at the origin, point O, and travels along a straight line path at a constant rate. Fifteen minutes into its journey the robot is at $A(35, 80)$.

   a. If the robot does not change speed or direction, where will it be 3 hours into its journey (call this point B)?

   b. The robot continues past point B for a certain period of time until it has traveled an additional $\dfrac{3}{4}$ of the distance it traveled in the first 3 hours and stops.

      i. How long did the robot's entire journey take?

      ii. What is the robot's final location?

      iii. What was the distance the robot traveled in the last leg of its journey?

5. Given $\overline{LM}$ and point R that lies on $\overline{LM}$, identify the following ratios given that point R lies $\dfrac{a}{b}$ of the way along $\overline{LM}$, closer to L than to M.

   a. $LR : LM$

   b. $RM : LM$

   c. $RL : RM$

6. Given $\overline{AB}$ with midpoint M as shown, prove that the point on the directed segment from A to B that divides $\overline{AB}$ into a ratio of $1 : 3$ is the midpoint of $\overline{AM}$.

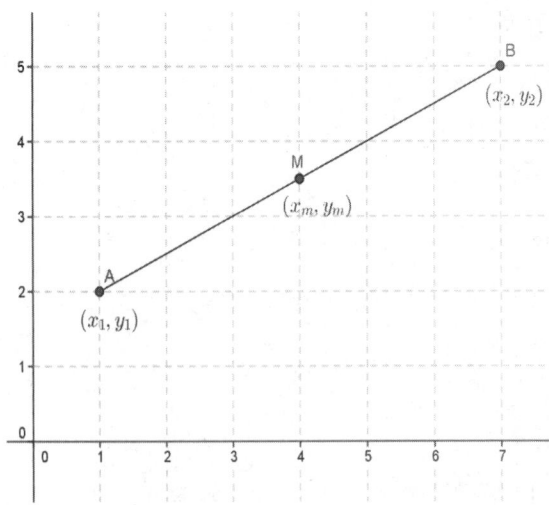

EUREKA
MATH™

© 2015 Great Minds. eureka-math.org
GEO-M3-SE-B2-1.3.0-10.2015

Lesson 13: Analytic Proofs of Theorems Previously Proved by Synthetic Means

Classwork

Opening Exercise

Let $A(30,40)$, $B(60,50)$, and $C(75,120)$ be vertices of a triangle.

a.   Find the coordinates of the midpoint M of $\overline{AB}$ and the point G_1 that is the point one-third of the way along $\overline{MC}$, closer to M than to C.

b.   Find the coordinates of the midpoint N of $\overline{BC}$ and the point G_2 that is the point one-third of the way along $\overline{NA}$, closer to N than to A.

c.   Find the coordinates of the midpoint R of $\overline{CA}$ and the point G_3 that is the point one-third of the way along $\overline{RB}$, closer to R than to B.

© 2015 Great Minds. eureka-math.org
GEO-M3-SE-B2-1.3.0-10.2015

Exercise 1

a. Given triangle ABC with vertices $A(a_1, a_2)$, $B(b_1, b_2)$, and $C(c_1, c_2)$, find the coordinates of the point of concurrency of the medians.

b. Let $A(-23, 12)$, $B(13, 36)$, and $C(23, -1)$ be vertices of a triangle. Where will the medians of this triangle intersect? (Use "Tyler's formula" from part (a) to complete this problem.)

© 2015 Great Minds. eureka-math.org
GEO-M3-SE-B2-1.3.0-10.2015

EUREKA MATH™

Exercise 2

Prove that the diagonals of a parallelogram bisect each other.

Problem Set

1. Point M is the midpoint of $\overline{AC}$. Find the coordinates of M:

    a. $A(2,3), C(6,10)$

    b. $A(-7,5), C(4,-9)$

2. $M(-2,10)$ is the midpoint of $\overline{AB}$. If A has coordinates $(4,-5)$, what are the coordinates of B?

3. Line A is the perpendicular bisector of $\overline{BC}$ with $B(-2,-1)$ and $C(4,1)$.

    a. What is the midpoint of $\overline{BC}$?

    b. What is the slope of $\overline{BC}$?

    c. What is the slope of line A? (Remember, it is perpendicular to $\overline{BC}$.)

    d. Write the equation of line A, the perpendicular bisector of $\overline{BC}$.

4. Find the coordinates of the intersection of the medians of $\triangle ABC$ given $A(-5,3)$, $B(6,-4)$, and $C(10,10)$.

5. Use coordinates to prove that the diagonals of a parallelogram meet at the intersection of the segments that connect the midpoints of its opposite sides.

6. Given a quadrilateral with vertices $E(0,5)$, $F(6,5)$, $G(4,0)$, and $H(-2,0)$:

    a. Prove quadrilateral $EFGH$ is a parallelogram.

    b. Prove $(2,2.5)$ is a point on both diagonals of the quadrilateral.

7. Prove quadrilateral $WXYZ$ with vertices $W(1,3)$, $X(4,8)$, $Y(10,11)$, and $Z(4,1)$ is a trapezoid.

8. Given quadrilateral $JKLM$ with vertices $J(-4,2)$, $K(1,5)$, $L(4,0)$, and $M(-1,-3)$:

    a. Is it a trapezoid? Explain.

    b. Is it a parallelogram? Explain.

    c. Is it a rectangle? Explain.

    d. Is it a rhombus? Explain.

    e. Is it a square? Explain.

    f. Name a point on the diagonal of $JKLM$. Explain how you know.

Lesson 14: Motion Along a Line—Search Robots Again

Opening Exercise

a. If $f(t) = (t, 2t - 1)$, find the values of $f(0)$, $f(1)$, and $f(5)$, and plot them on a coordinate plane.

b. What is the image of $f(t)$?

c. At what time does the graph of the line pass through the y-axis?

d. When does it pass through the x-axis?

e. Can you write the equation of the line you graphed in slope y-intercept form?

f. How does this equation compare with the definition of $f(t)$?

Example 1

Programmers want to program a robot so that it moves at a uniform speed along a straight line segment connecting two points A and B. If $A(0, -1)$ and $B(1, 1)$, and the robot travels from A to B in 1 minute,

a.   Where is the robot at $t = 0$?

b.   Where is the robot at $t = 1$?

c.   Draw a picture that shows where the robot will be at $0 \le t \le 1$.

Exercise 1

A robot is programmed to move along a straight line path through two points A and B. It travels at a uniform speed that allows it to make the trip from $A(0, -1)$ to $B(1, 1)$ in 1 minute. Find the robot's location, P, for each time t in minutes.

a. $t = \dfrac{1}{4}$

b. $t = 0.7$

EUREKA
MATH™

c. $t = \dfrac{5}{4}$

d. $t = 2.2$

Example 2

Our robot has been reprogrammed so that it moves along the same straight line path through two points $A(0, -1)$ and $B(1, 1)$ at a uniform rate but makes the trip in 0.6 minutes instead of 1 minute.

How does this change the way we calculate the location of the robot at any time, t?

a. Find the location, P, of the robot from Example 1 if the robot were traveling at a uniform speed that allowed it to make the trip from A to B in 0.6 minutes. Is the robot's speed greater or less than the robot's speed in Example 1?

b. Find the location, P, of the robot from Example 1 if the robot were traveling at a uniform speed that allowed it to make the trip from A to B in 1.5 minutes. Is the robot's speed greater or less than the robot's speed in Example 1?

Exercise 2

Two robots are moving along straight line paths in a rectangular room. Robot 1 starts at point $A(20, 10)$ and travels at a constant speed to point $B(120, 50)$ in 2 minutes. Robot 2 starts at point $C(90, 10)$ and travels at a constant speed to point $D(60, 70)$ in 90 seconds.

    a.   Find the location, P, of Robot 1 after it has traveled for t minutes along its path from A to B.

    b.   Find the location, Q, of Robot 2 after it has traveled for t minutes along its path from C to D.

    c.   Are the robots traveling at the same speed? If not, which robot's speed is greater?

    d.   Are the straight line paths that the robots are traveling parallel, perpendicular, or neither? Explain your answer.

© 2015 Great Minds. eureka-math.org
GEO-M3-SE-B2-1.3.0-10.2015

Example 3

A programmer wants to program a robot so that it moves at a constant speed along a straight line segment connecting the point $A(30, 60)$ to the point $B(200, 100)$ over the course of a minute.

At time $t = 0$, the robot is at point A.

At time $t = 1$, the robot is at point B.

   a.   Where will the robot be at time $t = \frac{1}{2}$?

   b.   Where will the robot be at time $t = 0.6$?

Problem Set

1. Find the coordinates of the intersection of the medians of $\triangle ABC$ given $A(2, 4)$, $B(-4, 0)$, and $C(3, -1)$.

2. Given a quadrilateral with vertices $A(-1, 3)$, $B(1, 5)$, $C(5, 1)$, and $D(3, -1)$:
   a. Prove that quadrilateral $ABCD$ is a rectangle.
   b. Prove that $(2, 2)$ is a point on both diagonals of the quadrilateral.

3. The robot is programed to travel along a line segment at a constant speed. If P represents the robot's position at any given time t in minutes:

$$P = (240, 60) + \frac{t}{10}(100, 100),$$

   a. What was the robot's starting position?
   b. Where did the robot stop?
   c. How long did it take the robot to complete the entire journey?
   d. Did the robot pass through the point $(310, 130)$, and, if so, how long into its journey did the robot reach this position?

4. Two robots are moving along straight line paths in a rectangular room. Robot 1 starts at point $A(20, 10)$ and travels at a constant speed to point $B(120, 50)$ in two minutes. Robot 2 starts at point $C(90, 10)$ and travels at a constant speed to point $D(60, 70)$ in 90 seconds. If the robots begin their journeys at the same time, will the robots collide? Why or why not?

Lesson 15: The Distance from a Point to a Line

Classwork

Exercise 1

A robot is moving along the line $20x + 30y = 600$. A homing beacon sits at the point $(35, 40)$.

    a.   Where on this line will the robot hear the loudest ping?

    b.   At this point, how far will the robot be from the beacon?

Exercise 2

For the following problems, use the formula to calculate the distance between the point P and the line l.

$$d = \sqrt{\left(\frac{p+qm-bm}{1+m^2}-p\right)^2 + \left(m\left(\frac{p+qm-bm}{1+m^2}\right)+b-q\right)^2}$$

a. $P(0,0)$ and the line $y = 10$

b. $P(0,0)$ and the line $y = x + 10$

c. $P(0,0)$ and the line $y = x - 6$

EUREKA
MATH™

© 2015 Great Minds. eureka-math.org
GEO-M3-SE-B2-1.3.0-10.2015

Problem Set

1. Given $\triangle ABC$ with vertices $A(3, -1)$, $B(2, 2)$, and $C(5, 1)$.

   a. Find the slope of the angle bisector of $\angle ABC$.

   b. Prove that the bisector of $\angle ABC$ is the perpendicular bisector of $\overline{AC}$.

   c. Write the equation of the line containing $\overline{BD}$, where point D is the point where the bisector of $\angle ABC$ intersects $\overline{AC}$.

2. Use the distance formula from today's lesson to find the distance between the point $P(-2, 1)$ and the line $y = 2x$.

3. Confirm the results obtained in Problem 2 using another method.

4. Find the perimeter of quadrilateral $DEBF$ shown below.

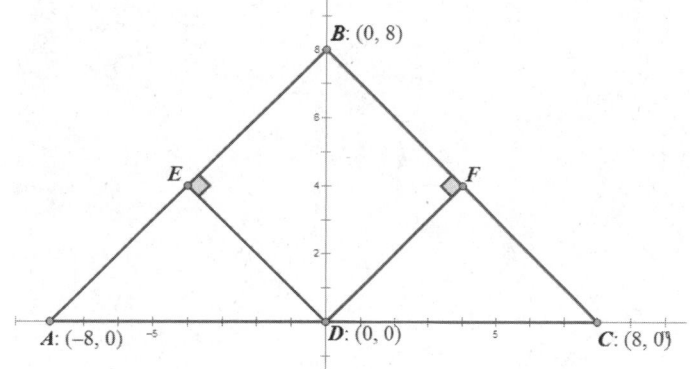

EUREKA
MATH™

Lesson 15:   The Distance from a Point to a Line

S.75

© 2015 Great Minds. eureka-math.org
GEO-M3-SE-B2-1.3.0-10.2015

This page intentionally left blank

Eureka Math
Geometry
Module 5

Special thanks go to the Gordon A. Cain Center and to the Department of Mathematics at Louisiana State University for their support in the development of *Eureka Math*.

For a free *Eureka Math* Teacher Resource Pack, Parent Tip Sheets, and more please visit www.Eureka.tools

Published by the non-profit Great Minds

Copyright © 2015 Great Minds. No part of this work may be reproduced, sold, or commercialized, in whole or in part, without written permission from Great Minds. Non-commercial use is licensed pursuant to a Creative Commons Attribution-NonCommercial-ShareAlike 4.0 license; for more information, go to http://greatminds.net/maps/math/copyright. "Great Minds" and "Eureka Math" are registered trademarks of Great Minds.

Printed in the U.S.A.
This book may be purchased from the publisher at eureka-math.org
1 2 3 4 5 6 7 8   BAB   25 24 23 22 21

ISBN 978-1-63255-328-7

Lesson 1: Thales' Theorem

Opening Exercise

a.   Mark points A and B on the sheet of white paper provided by your teacher.

b.   Take the colored paper provided, and *push* that paper up between points A and B on the white sheet.

c.   Mark on the white paper the location of the corner of the colored paper, using a different color than black. Mark that point C. See the example below.

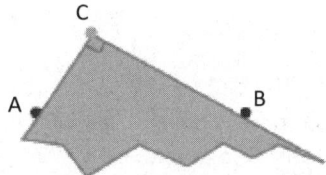

d.   Do this again, pushing the corner of the colored paper up between the black points but at a different angle. Again, mark the location of the corner. Mark this point D.

e.   Do this again and then again, multiple times. Continue to label the points. What curve do the colored points $(C, D, …)$ seem to trace?

Exploratory Challenge

Choose one of the colored points $(C, D, …)$ that you marked. Draw the right triangle formed by the line segment connecting the original two points A and B and that colored point. Take a copy of the triangle, and rotate it $180°$ about the midpoint of $\overline{AB}$.

Label the acute angles in the original triangle as x and y, and label the corresponding angles in the rotated triangle the same.

Todd says $ACBC'$ is a rectangle. Maryam says $ACBC'$ is a quadrilateral, but she is not sure it is a rectangle. Todd is right but does not know how to explain himself to Maryam. Can you help him out?

a.   What composite figure is formed by the two triangles? How would you prove it?

i.   What is the sum of the measures of x and y? Why?

     ii.   How do we know that the figure whose vertices are the colored points (C, D, …) and points A and B is a rectangle?

    b.   Draw the two diagonals of the rectangle. Where is the midpoint of the segment connecting the two original points A and B? Why?

    c.   Label the intersection of the diagonals as point P. How does the distance from point P to a colored point (C, D, …) compare to the distance from P to points A and B?

    d.   Choose another colored point, and construct a rectangle using the same process you followed before. Draw the two diagonals of the new rectangle. How do the diagonals of the new and old rectangle compare? How do you know?

    e.   How does your drawing demonstrate that all the colored points you marked do indeed lie on a circle?

        Lesson 1:   Thales' Theorem

EUREKA MATH™

Example

In the Exploratory Challenge, you proved the converse of a famous theorem in geometry. Thales' theorem states the following: *If A, B, and C are three distinct points on a circle, and $\overline{AB}$ is a diameter of the circle, then $\angle ACB$ is right.*

Notice that, in the proof in the Exploratory Challenge, you started with a right angle (the corner of the colored paper) and created a circle. With Thales' theorem, you must start with the circle and then create a right angle.

Prove Thales' theorem.

    a.   Draw circle P with distinct points A, B, and C on the circle and diameter $\overline{AB}$. Prove that $\angle ACB$ is a right angle.

    b.   Draw a third radius ($\overline{PC}$). What types of triangles are $\triangle APC$ and $\triangle BPC$? How do you know?

    c.   Using the diagram that you just created, develop a strategy to prove Thales' theorem.

    d.   Label the base angles of $\triangle APC$ as $b°$ and the base angles of $\triangle BPC$ as $a°$. Express the measure of $\angle ACB$ in terms of $a°$ and $b°$.

    e.   How can the previous conclusion be used to prove that $\angle ACB$ is a right angle?

Lesson 1:   Thales' Theorem   S.3

© 2015 Great Minds. eureka-math.org
GEO-M3-SE-B2-1.3.0-10.2015

Exercises

1. $\overline{AB}$ is a diameter of the circle shown. The radius is 12.5 cm, and $AC = 7$ cm.

   a. Find $m\angle C$.

   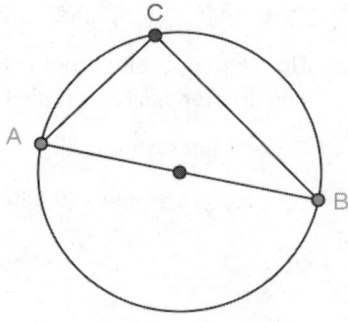

   b. Find AB.

   c. Find BC.

2. In the circle shown, $\overline{BC}$ is a diameter with center A.

   a. Find $m\angle DAB$.

   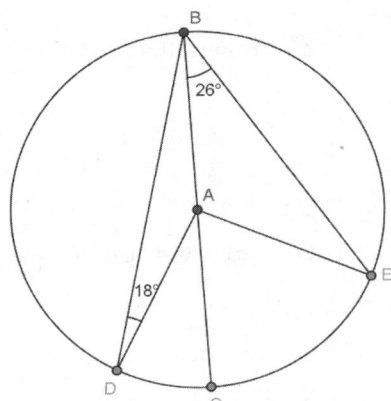

   b. Find $m\angle BAE$.

   c. Find $m\angle DAE$.

EUREKA
MATH™

Lesson Summary

Theorems:

- THALES' THEOREM: If A, B, and C are three different points on a circle with a diameter $\overline{AB}$, then $\angle ACB$ is a right angle.

- CONVERSE OF THALES' THEOREM: If $\triangle ABC$ is a right triangle with $\angle C$ the right angle, then A, B, and C are three distinct points on a circle with a diameter $\overline{AB}$.

- Therefore, given distinct points A, B, and C on a circle, $\triangle ABC$ is a right triangle with $\angle C$ the right angle if and only if $\overline{AB}$ is a diameter of the circle.

- Given two points A and B, let point P be the midpoint between them. If C is a point such that $\angle ACB$ is right, then $BP = AP = CP$.

Relevant Vocabulary

- CIRCLE: Given a point C in the plane and a number $r > 0$, the *circle* with center C and radius r is the set of all points in the plane that are distance r from the point C.

- RADIUS: May refer either to the line segment joining the center of a circle with any point on that circle (a *radius*) or to the length of this line segment (the *radius*).

- DIAMETER: May refer either to the segment that passes through the center of a circle whose endpoints lie on the circle (a *diameter*) or to the length of this line segment (the *diameter*).

- CHORD: Given a circle C, and let P and Q be points on C. $\overline{PQ}$ is called a *chord* of C.

- CENTRAL ANGLE: A *central angle* of a circle is an angle whose vertex is the center of a circle.

Problem Set

1. A, B, and C are three points on a circle, and angle ABC is a right angle. What is wrong with the picture below? Explain your reasoning.

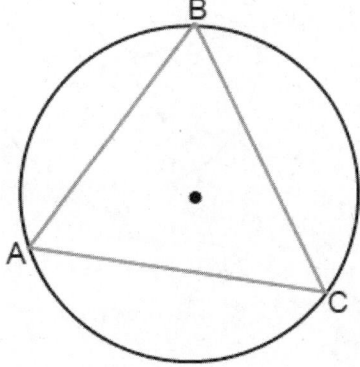

2. Show that there is something mathematically wrong with the picture below.

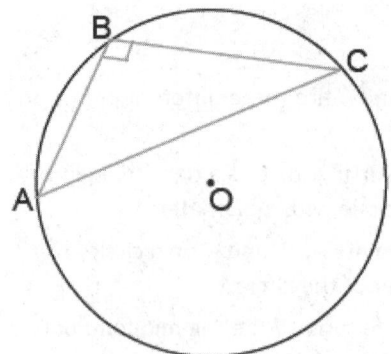

3. In the figure below, $\overline{AB}$ is the diameter of a circle of radius 17 miles. If $BC = 30$ miles, what is AC?

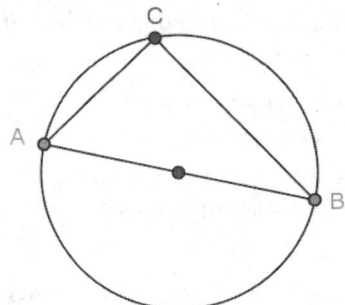

4. In the figure below, O is the center of the circle, and $\overline{AD}$ is a diameter.

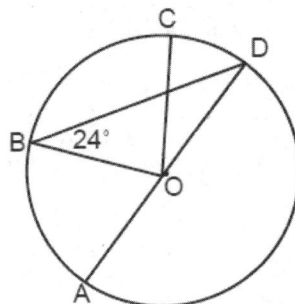

    a. Find $m\angle AOB$.

    b. If $m\angle AOB : m\angle COD = 3 : 4$, what is $m\angle BOC$?

EUREKA
MATH™

5. $\overline{PQ}$ is a diameter of a circle, and M is another point on the circle. The point R lies on $\overleftrightarrow{MQ}$ such that $RM = MQ$. Show that $m\angle PRM = m\angle PQM$. (Hint: Draw a picture to help you explain your thinking.)

6. Inscribe $\triangle ABC$ in a circle of diameter 1 such that $\overline{AC}$ is a diameter. Explain why:

   a. $\sin(\angle A) = BC$.

   b. $\cos(\angle A) = AB$.

This page intentionally left blank

Lesson 2: Circles, Chords, Diameters, and Their Relationships

Classwork

Opening Exercise

Construct the perpendicular bisector of $\overline{AB}$ below (as you did in Module 1).

A ——————————— B

Draw another line that bisects $\overline{AB}$ but is not perpendicular to it.

List one similarity and one difference between the two bisectors.

Exercises

Figures are not drawn to scale.

1. Prove the theorem: *If a diameter of a circle bisects a chord, then it must be perpendicular to the chord.*

2. Prove the theorem: *If a diameter of a circle is perpendicular to a chord, then it bisects the chord.*

Lesson 2:   Circles, Chords, Diameters, and Their Relationships

© 2015 Great Minds. eureka-math.org
GEO-M3-SE-B2-1.3.0-10.2015

3. The distance from the center of a circle to a chord is defined as the length of the perpendicular segment from the center to the chord. Note that since this perpendicular segment may be extended to create a diameter of the circle, the segment also bisects the chord, as proved in Exercise 2.

    Prove the theorem: *In a circle, if two chords are congruent, then the center is equidistant from the two chords.*

    Use the diagram below.

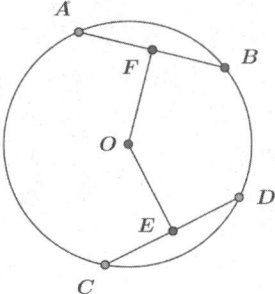

4. Prove the theorem: *In a circle, if the center is equidistant from two chords, then the two chords are congruent.*

Use the diagram below.

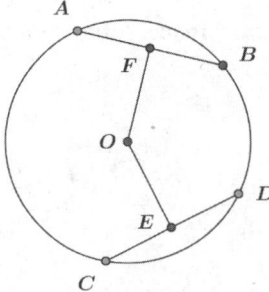

Lesson 2:   Circles, Chords, Diameters, and Their Relationships

EUREKA
MATH™

5. A central angle defined by a chord is an angle whose vertex is the center of the circle and whose rays intersect the circle. The points at which the angle's rays intersect the circle form the endpoints of the chord defined by the central angle.

    Prove the theorem: *In a circle, congruent chords define central angles equal in measure.*

    Use the diagram below.

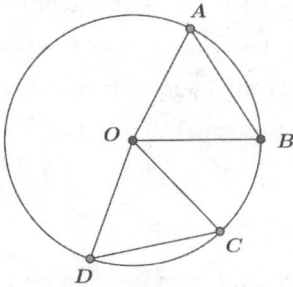

6. Prove the theorem: *In a circle, if two chords define central angles equal in measure, then they are congruent.*

Problem Set

Figures are not drawn to scale.

1. In this drawing, $AB = 30$, $OM = 20$, and $ON = 18$. What is CN?

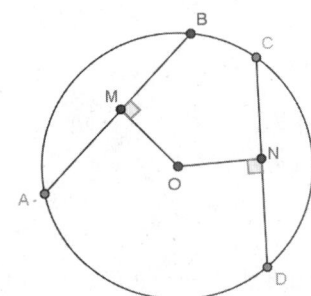

2. In the figure to the right, $\overline{AC} \perp \overline{BG}$, $\overline{DF} \perp \overline{EG}$, and $EF = 12$. Find AC.

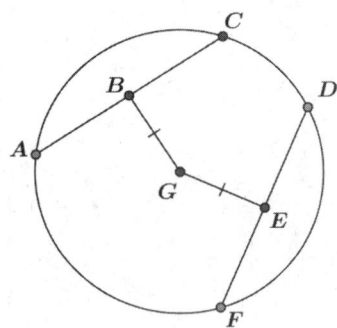

EUREKA
MATH™

3. In the figure, $AC = 24$, and $DG = 13$. Find EG. Explain your work.

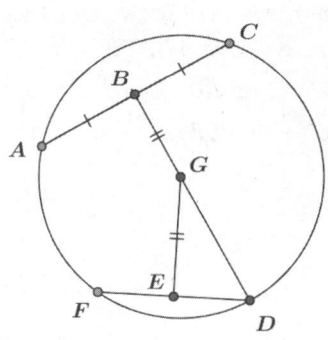

4. In the figure, $AB = 10$, and $AC = 16$. Find DE.

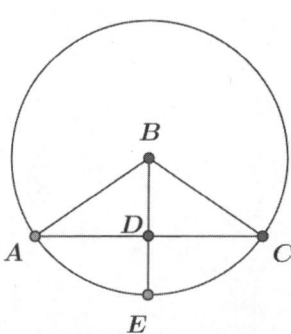

5. In the figure, $CF = 8$, and the two concentric circles have radii of 10 and 17. Find DE.

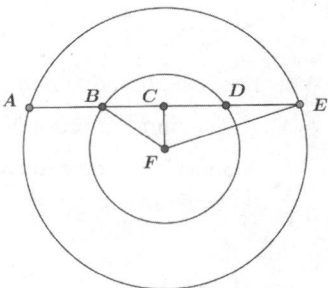

6. In the figure, the two circles have equal radii and intersect at points B and D. A and C are centers of the circles. $AC = 8$, and the radius of each circle is 5. $\overline{BD} \perp \overline{AC}$. Find BD. Explain your work.

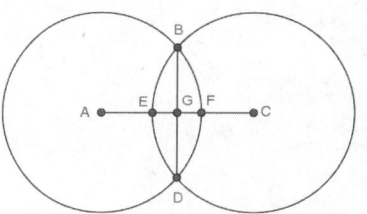

EUREKA
MATH™

Lesson 2:   Circles, Chords, Diameters, and Their Relationships

S.15

© 2015 Great Minds. eureka-math.org
GEO-M3-SE-B2-1.3.0-10.2015

7. In the figure, the two concentric circles have radii of 6 and 14. Chord $\overline{BF}$ of the larger circle intersects the smaller circle at C and E. $CE = 8$. $\overline{AD} \perp \overline{BF}$.

   a. Find AD.

   b. Find BF.

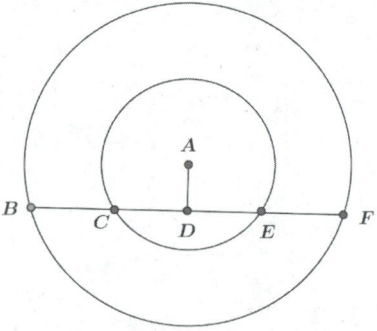

8. In the figure, A is the center of the circle, and $CB = CD$. Prove that $\overline{AC}$ bisects $\angle BCD$.

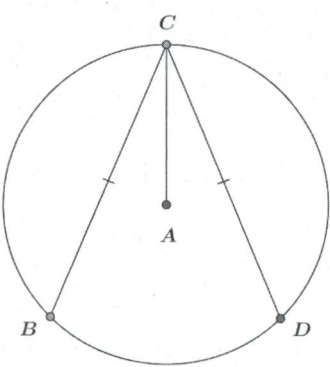

9. In class, we proved: *Congruent chords define central angles equal in measure.*

   a. Give another proof of this theorem based on the properties of rotations. Use the figure from Exercise 5.

   b. Give a rotation proof of the converse: *If two chords define central angles of the same measure, then they must be congruent.*

EUREKA
MATH

Graphic Organizer on Circles

Diagram	Explanation of Diagram	Theorem or Relationship

This page intentionally left blank

Lesson 3: Rectangles Inscribed in Circles

Classwork

Opening Exercise

Using only a compass and straightedge, find the location of the center of the circle below. Follow the steps provided.

- Draw chord $\overline{AB}$.
- Construct a chord perpendicular to $\overline{AB}$ at endpoint B.
- Mark the point of intersection of the perpendicular chord and the circle as point C.
- $\overline{AC}$ is a diameter of the circle. Construct a second diameter in the same way.
- Where the two diameters meet is the center of the circle.

Explain why the steps of this construction work.

Exploratory Challenge

Construct a rectangle such that all four vertices of the rectangle lie on the circle below.

Exercises

1. Construct a kite inscribed in the circle below, and explain the construction using symmetry.

2. Given a circle and a rectangle, what must be true about the rectangle for it to be possible to inscribe a congruent copy of it in the circle?

3. The figure below shows a rectangle inscribed in a circle.

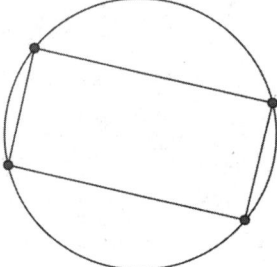

   a. List the properties of a rectangle.

b.   List all the symmetries this diagram possesses.

c.   List the properties of a square.

d.   List all the symmetries of the diagram of a square inscribed in a circle.

4.   A rectangle is inscribed into a circle. The rectangle is cut along one of its diagonals and reflected across that diagonal to form a kite. Draw the kite and its diagonals. Find all the angles in this new diagram, given that the acute angle formed by the diagonal of the rectangle in the original diagram was 40°.

5. **Challenge:** Show that the three vertices of a right triangle are equidistant from the midpoint of the hypotenuse by showing that the perpendicular bisectors of the legs pass through the midpoint of the hypotenuse.

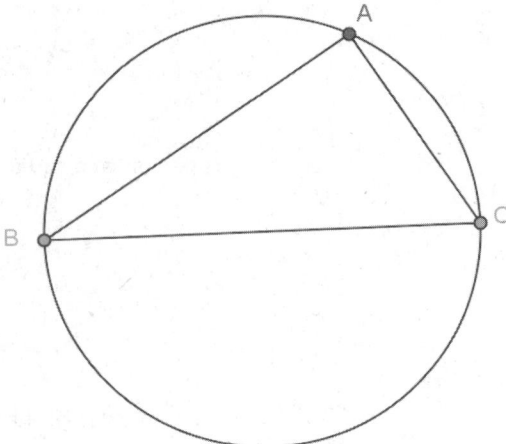

a. Draw the perpendicular bisectors of $\overline{AB}$ and $\overline{AC}$.

b. Label the point where they meet P. What is point P?

c. What can be said about the distance from P to each vertex of the triangle? What is the relationship between the circle and the triangle?

d. Repeat this process, this time sliding B to another place on the circle and call it B'. What do you notice?

e. Is there a relationship between $m\angle ABC$ and $m\angle AB'C$? Explain.

EUREKA
MATH™

Lesson Summary

Relevant Vocabulary

INSCRIBED POLYGON: A polygon is *inscribed* in a circle if all vertices of the polygon lie on the circle.

Problem Set

Figures are not drawn to scale.

1. Using only a piece of 8.5×11 inch copy paper and a pencil, find the location of the center of the circle below.

2. Is it possible to inscribe a parallelogram that is not a rectangle in a circle?

3. In the figure, $BCDE$ is a rectangle inscribed in circle A. $DE = 8$; $BE = 12$. Find AE.

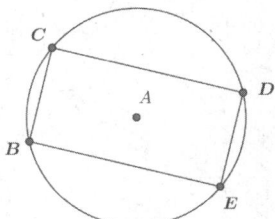

4. Given the figure, $BC = CD = 8$ and $AD = 13$. Find the radius of the circle.

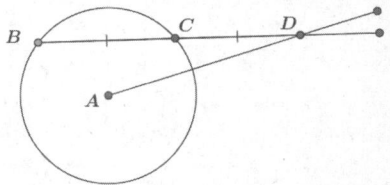

5. In the figure, $\overline{DF}$ and $\overline{BG}$ are parallel chords 14 cm apart. $DF = 12$ cm, $AB = 10$ cm, and $\overline{EH} \perp \overline{BG}$. Find BG.

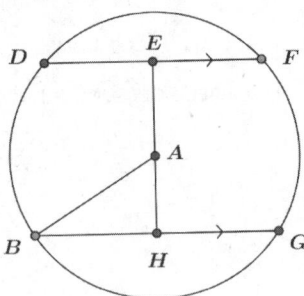

6. Use perpendicular bisectors of the sides of a triangle to construct a circle that circumscribes the triangle.

EUREKA
MATH™

Lesson 4: Experiments with Inscribed Angles

Opening Exercise

ARC:

MINOR AND MAJOR ARC:

INSCRIBED ANGLE:

CENTRAL ANGLE:

INTERCEPTED ARC OF AN ANGLE:

Exploratory Challenge 1

Your teacher will provide you with a straightedge, a sheet of colored paper in the shape of a trapezoid, and a sheet of plain white paper.

- Draw two points no more than 3 inches apart in the middle of the plain white paper, and label them A and B.
- Use the acute angle of your colored trapezoid to plot a point on the white sheet by placing the colored cutout so that the points A and B are on the edges of the acute angle and then plotting the position of the vertex of the angle. Label that vertex C.
- Repeat several times. Name the points $D, E, \ldots$.

Exploratory Challenge 2

a. Draw several of the angles formed by connecting points A and B on your paper with any of the additional points you marked as the acute angle was pushed through the points ($C, D, E, \ldots$). What do you notice about the measures of these angles?

b. Draw several of the angles formed by connecting points A and B on your paper with any of the additional points you marked as the obtuse angle was pushed through the points from above. What do you notice about the measures of these angles?

Exploratory Challenge 3

a. Draw a point on the circle, and label it D. Create angle $\angle BDC$.

b. $\angle BDC$ is called an inscribed angle. Can you explain why?

c.   Arc BC is called the intercepted arc. Can you explain why?

d.   Carefully cut out the inscribed angle, and compare it to the angles of several of your neighbors.

e.   What appears to be true about each of the angles you drew?

f.   Draw another point on a second circle, and label it point E. Create $\angle BEC$, and cut it out. Compare $\angle BDC$ and $\angle BEC$. What appears to be true about the two angles?

g.   What conclusion may be drawn from this? Will all angles inscribed in the circle from these two points have the same measure?

h.   Explain to your neighbor what you have just discovered.

Exploratory Challenge 4

a.   In the circle below, draw the angle formed by connecting points B and C to the center of the circle.

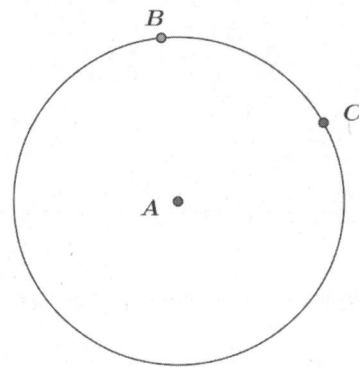

b.   Is $\angle BAC$ an inscribed angle? Explain.

c.   Is it appropriate to call this *the* central angle? Why or why not?

d.   What is the intercepted arc?

e.   Is the measure of $\angle BAC$ the same as the measure of one of the inscribed angles in Exploratory Challenge 2?

f.   Can you make a prediction about the relationship between the inscribed angle and the central angle?

Lesson Summary

All inscribed angles from the same intercepted arc have the same measure.

Relevant Vocabulary

- **ARC:** An *arc* is a portion of the circumference of a circle.

- **MINOR AND MAJOR ARC:** Let C be a circle with center O, and let A and B be different points that lie on C but are not the endpoints of the same diameter. The *minor arc* is the set containing A, B, and all points of C that are in the interior of $\angle AOB$. The *major arc* is the set containing A, B, and all points of C that lie in the exterior of $\angle AOB$.

- **INSCRIBED ANGLE:** An *inscribed angle* is an angle whose vertex is on a circle, and each side of the angle intersects the circle in another point.

- **CENTRAL ANGLE:** A *central angle* of a circle is an angle whose vertex is the center of a circle.

- **INTERCEPTED ARC OF AN ANGLE:** An angle *intercepts* an arc if the endpoints of the arc lie on the angle, all other points of the arc are in the interior of the angle, and each side of the angle contains an endpoint of the arc.

Problem Set

1. Using a protractor, measure both the inscribed angle and the central angle shown on the circle below.

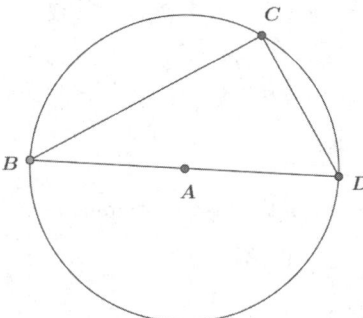

$m\angle BCD =$ _____ $m\angle BAD =$ _____

2. Using a protractor, measure both the inscribed angle and the central angle shown on the circle below.

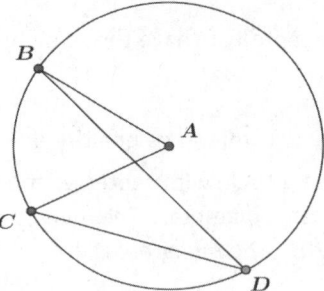

$m\angle BDC =$ _____ $m\angle BAC =$ _____

3. Using a protractor, measure both the inscribed angle and the central angle shown on the circle below.

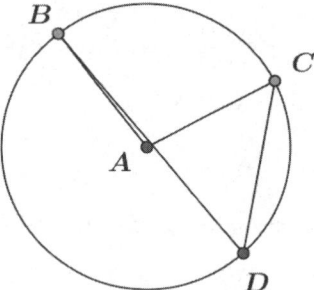

$m\angle BDC =$ _____ $m\angle BAC =$ _____

4. What relationship between the measure of the inscribed angle and the measure of the central angle that intercepts the same arc is illustrated by these examples?

5. Is your conjecture at least true for inscribed angles that measure 90°?

6. Prove that $y = 2x$ in the diagram below.

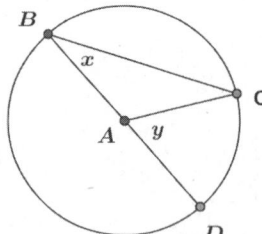

EUREKA
MATH™

7. Red (R) and blue (B) lighthouses are located on the coast of the ocean. Ships traveling are in safe waters as long as the angle from the ship (S) to the two lighthouses ($\angle RSB$) is always less than or equal to some angle θ called the *danger angle*. What happens to θ as the ship gets closer to shore and moves away from shore? Why do you think a larger angle is dangerous?

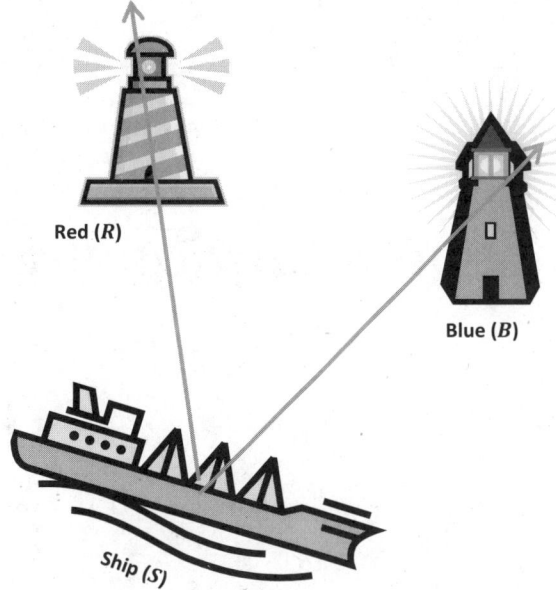

EUREKA
MATH™

This page intentionally left blank

Exploratory Challenge 3

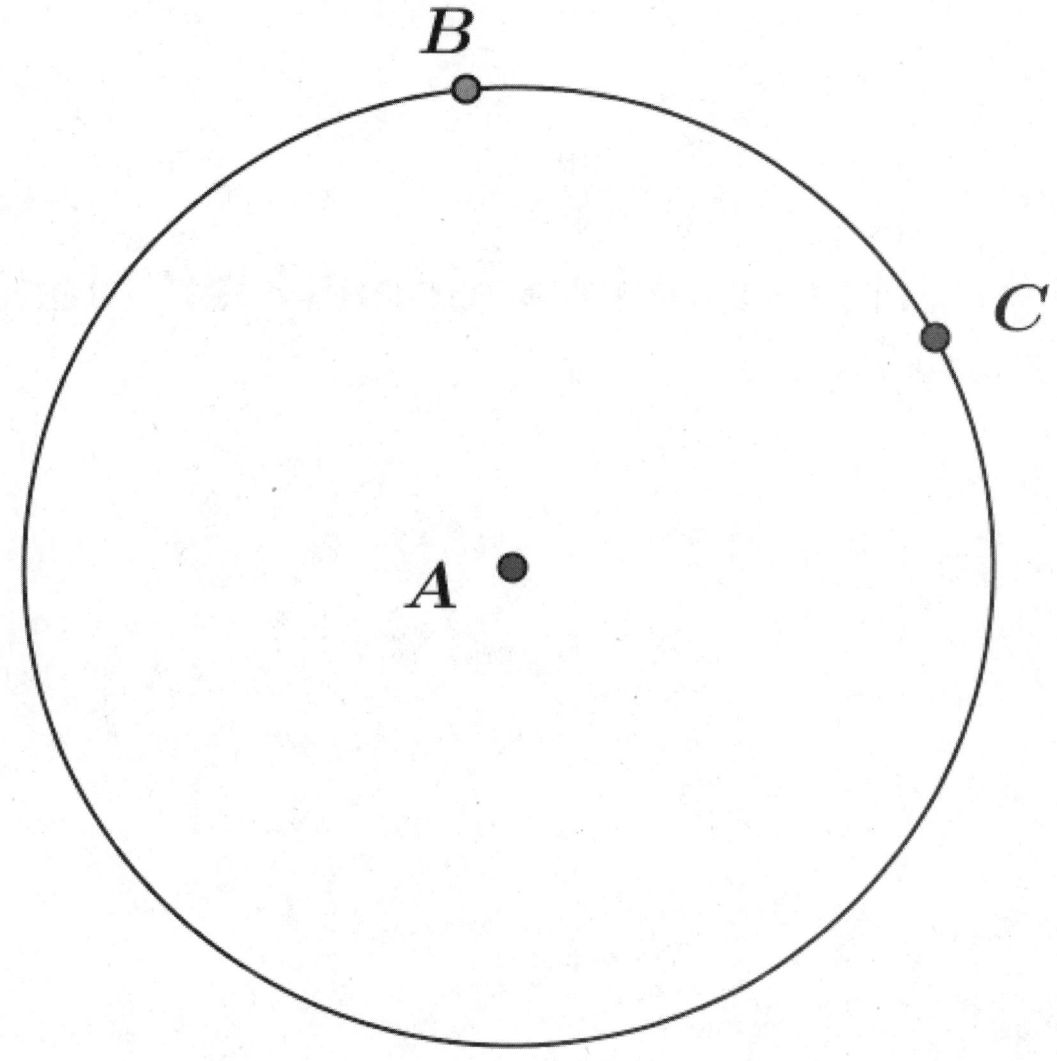

EUREKA
MATH™

This page intentionally left blank

Lesson 5: Inscribed Angle Theorem and Its Applications

Classwork

Opening Exercise

a. A and C are points on a circle with center O.

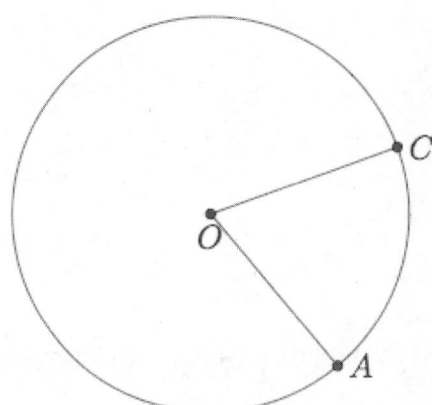

   i.   Draw a point B on the circle so that $\overline{AB}$ is a diameter. Then draw the angle ABC.

   ii.   What angle in your diagram is an inscribed angle?

   iii.   What angle in your diagram is a central angle?

   iv.   What is the intercepted arc of $\angle ABC$?

   v.   What is the intercepted arc of $\angle AOC$?

b.   The measure of the inscribed angle ACD is x, and the measure of the central angle CAB is y. Find $m\angle CAB$ in terms of x.

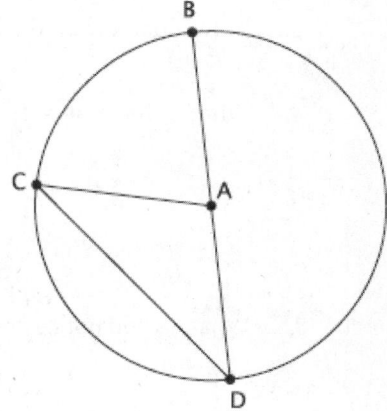

Example 1

A and C are points on a circle with center O.

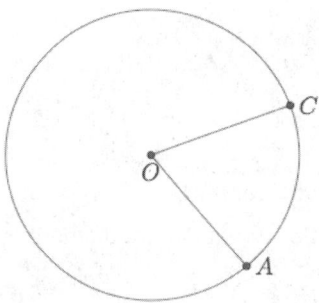

a.   What is the intercepted arc of $\angle COA$? Color it red.

b.   Draw triangle AOC. What type of triangle is it? Why?

c.   What can you conclude about $m\angle OCA$ and $m\angle OAC$? Why?

d.   Draw a point B on the circle so that O is <u>in the interior of</u> the inscribed angle ABC.

e.   What is the intercepted arc of $\angle ABC$? Color it green.

f.   What do you notice about $\overset{\frown}{AC}$?

EUREKA
MATH™

g. Let the measure of $\angle ABC$ be x and the measure of $\angle AOC$ be y. Can you prove that $y = 2x$? (Hint: Draw the diameter that contains point B.)

h. Does your conclusion support the inscribed angle theorem?

i. If we combine the Opening Exercise and this proof, have we finished proving the inscribed angle theorem?

Example 2

A and C are points on a circle with center O.

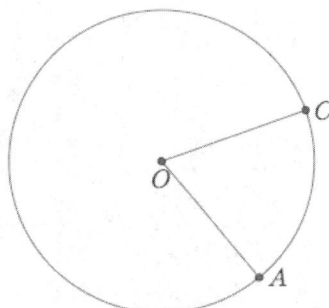

a. Draw a point B on the circle so that O is in the exterior of the inscribed angle ABC.

b. What is the intercepted arc of $\angle ABC$? Color it yellow.

c. Let the measure of $\angle ABC$ be x and the measure of $\angle AOC$ be y. Can you prove that $y = 2x$? (Hint: Draw the diameter that contains point B.)

d. Does your conclusion support the inscribed angle theorem?

e. Have we finished proving the inscribed angle theorem?

Exercises

1. Find the measure of the angle with measure x. Diagrams are not drawn to scale.

a. $m\angle D = 25°$

b. $m\angle D = 15°$

c. $m\angle BAC = 90°$

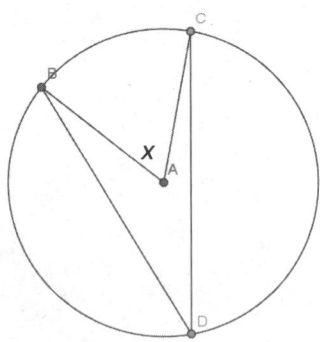

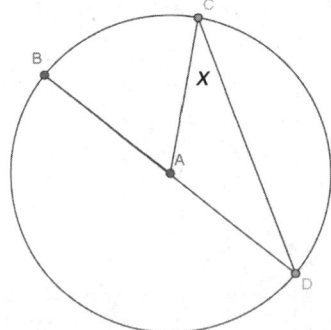

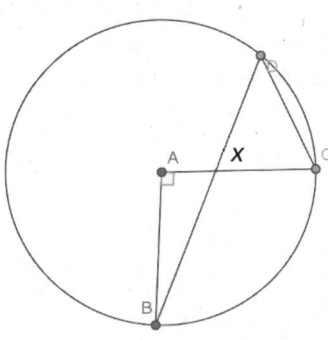

Lesson 5:   Inscribed Angle Theorem and Its Applications

EUREKA
MATH™

d. $m\angle B = 32°$

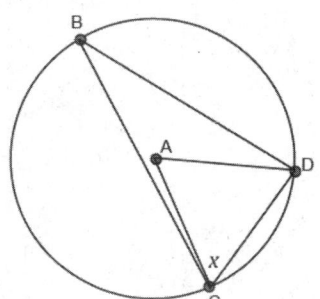

e. $BD = AB$

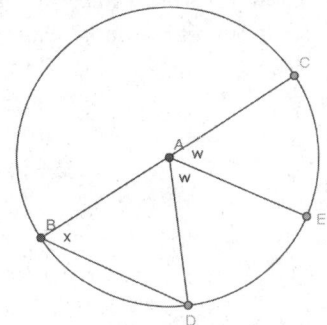

f. $m\angle D = 19°$

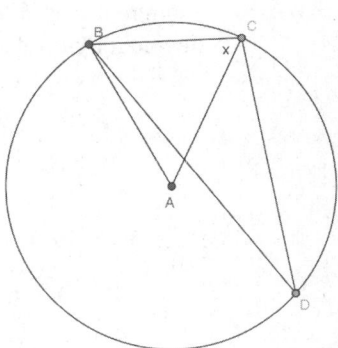

2.   Toby says $\triangle BEA$ is a right triangle because $m\angle BEA = 90°$. Is he correct? Justify your answer.

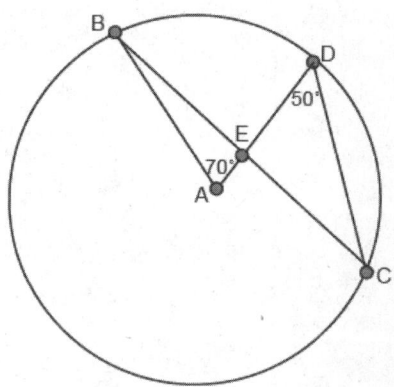

3. Let's look at relationships between inscribed angles.

    a. Examine the inscribed polygon below. Express x in terms of y and y in terms of x. Are the opposite angles in any quadrilateral inscribed in a circle supplementary? Explain.

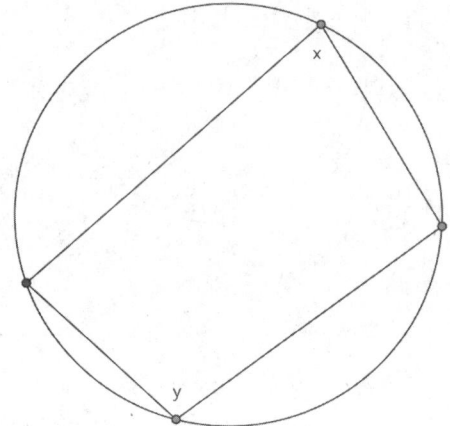

    b. Examine the diagram below. How many angles have the same measure, and what are their measures in terms of $x°$?

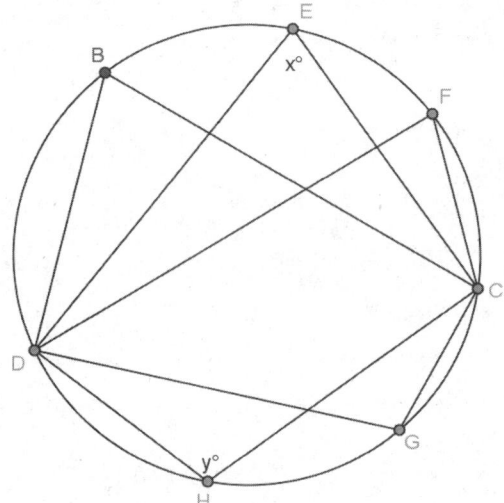

© 2015 Great Minds. eureka-math.org
GEO-M3-SE-B2-1.3.0-10.2015

EUREKA
MATH™

4. Find the measures of the labeled angles.

a.

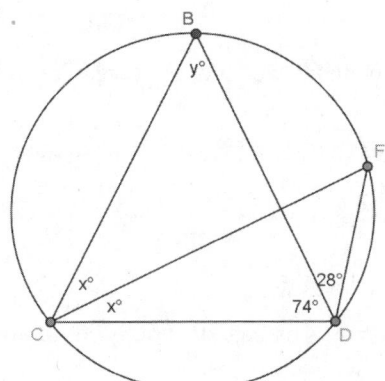

b.

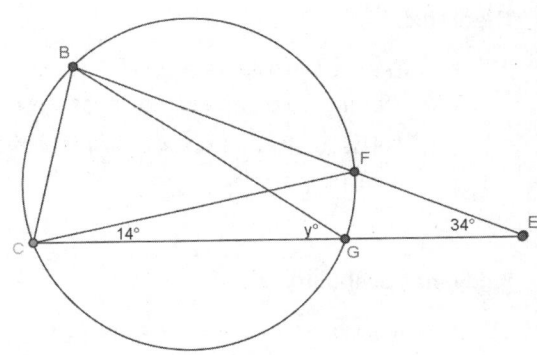

c.

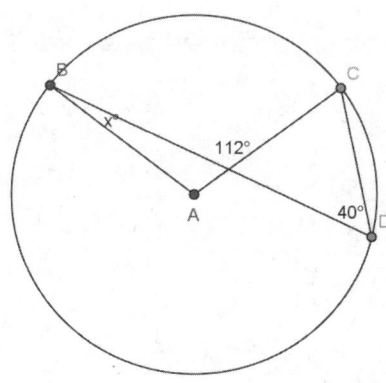

d.

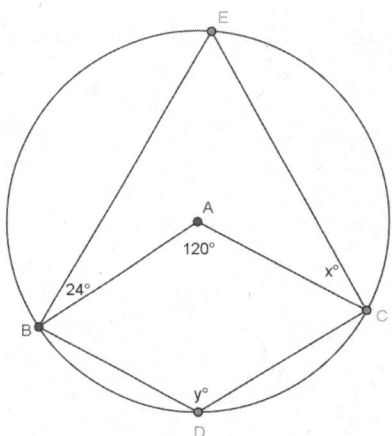

e.

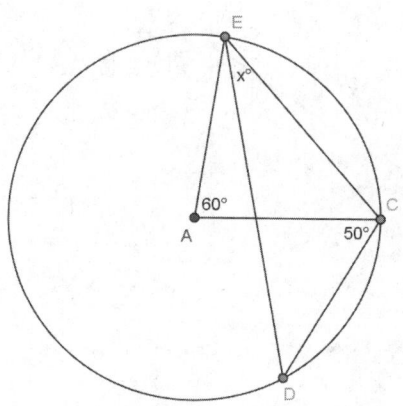

f.

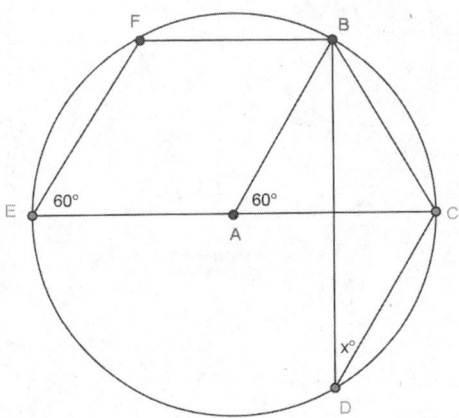

Lesson Summary

Theorems:

- **THE INSCRIBED ANGLE THEOREM:** The measure of a central angle is twice the measure of any inscribed angle that intercepts the same arc as the central angle.

- **CONSEQUENCE OF INSCRIBED ANGLE THEOREM:** Inscribed angles that intercept the same arc are equal in measure.

Relevant Vocabulary

- **INSCRIBED ANGLE:** An *inscribed angle* is an angle whose vertex is on a circle, and each side of the angle intersects the circle in another point.

- **INTERCEPTED ARC:** An angle *intercepts* an arc if the endpoints of the arc lie on the angle, all other points of the arc are in the interior of the angle, and each side of the angle contains an endpoint of the arc. An angle inscribed in a circle intercepts exactly one arc, in particular, the arc intercepted by an inscribed right angle is the semicircle in the interior of the angle.

Problem Set

For Problems 1–8, find the value of x. Diagrams are not drawn to scale.

1.

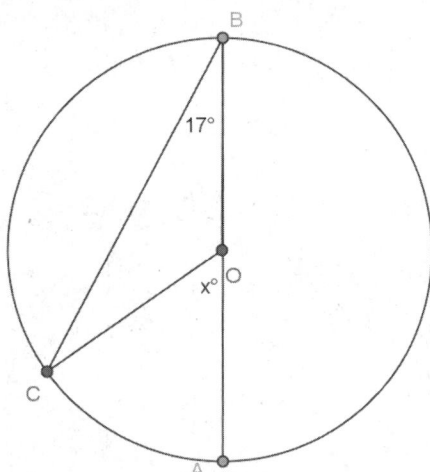

2.

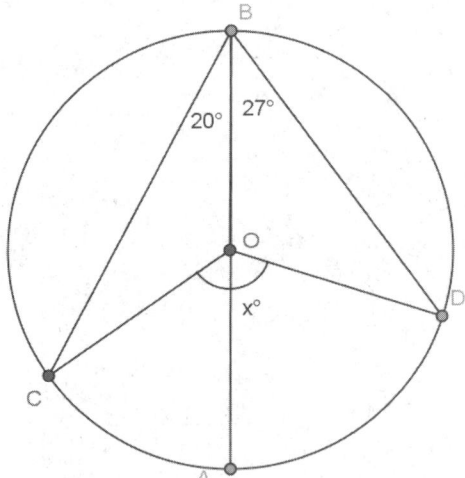

3.

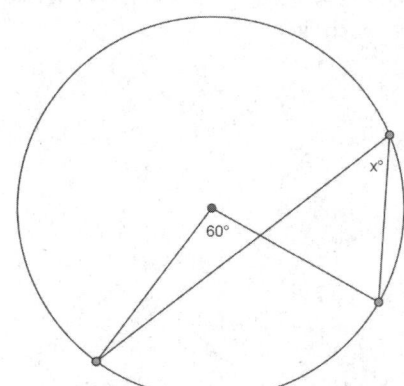

4.

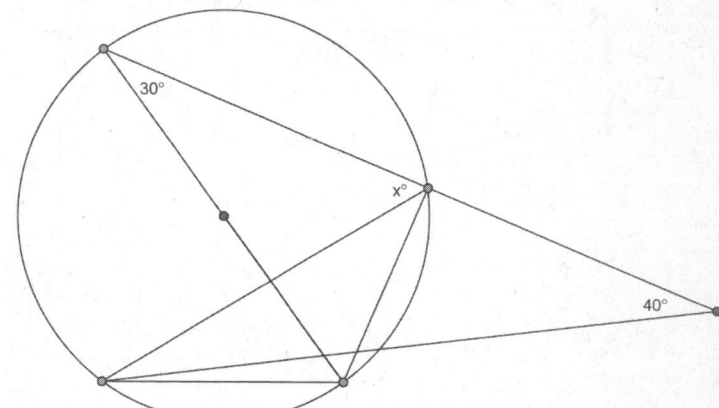

5.

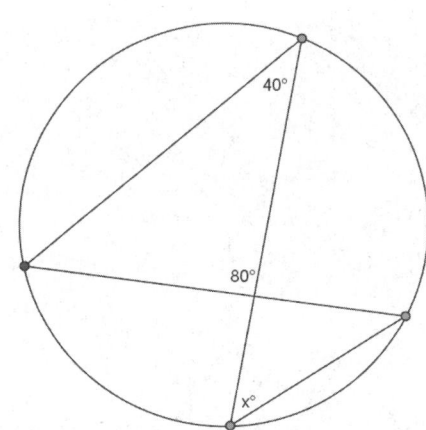

6.

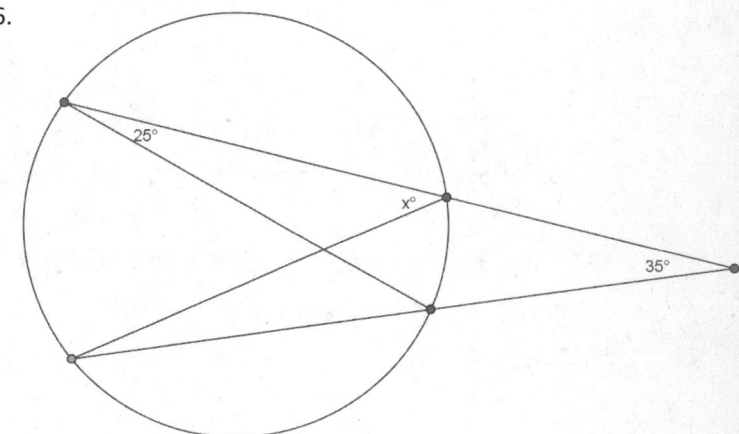

7.

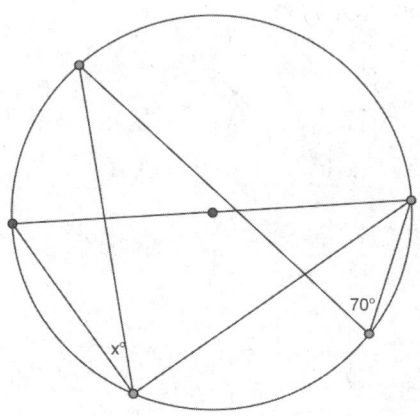

8.

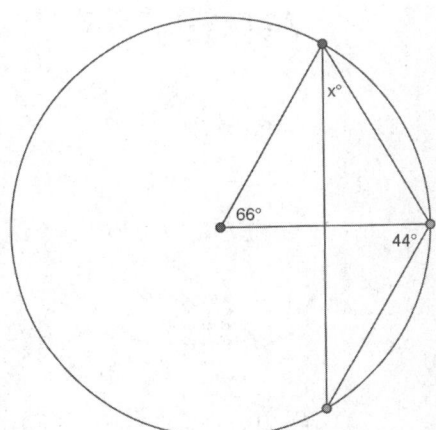

EUREKA
MATH™

9.

    a.   The two circles shown intersect at E and F. The center of the larger circle, D, lies on the circumference of the smaller circle. If a chord of the larger circle, $\overline{FG}$, cuts the smaller circle at H, find x and y.

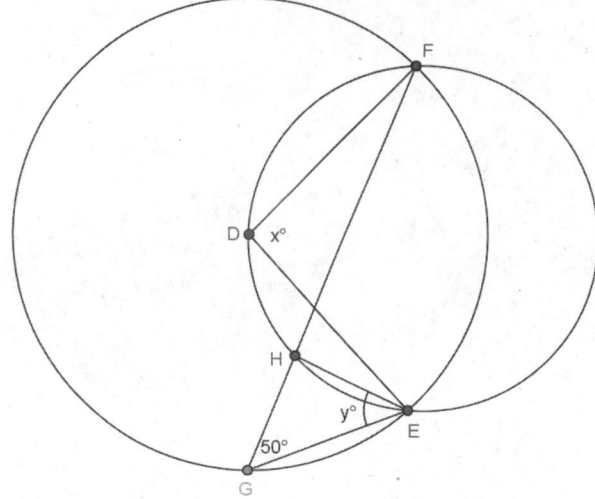

    b.   How does this problem confirm the inscribed angle theorem?

10. In the figure below, $\overline{ED}$ and $\overline{BC}$ intersect at point F.

    Prove: $m\angle DAB + m\angle EAC = 2(m\angle BFD)$

PROOF: Join $\overline{BE}$.

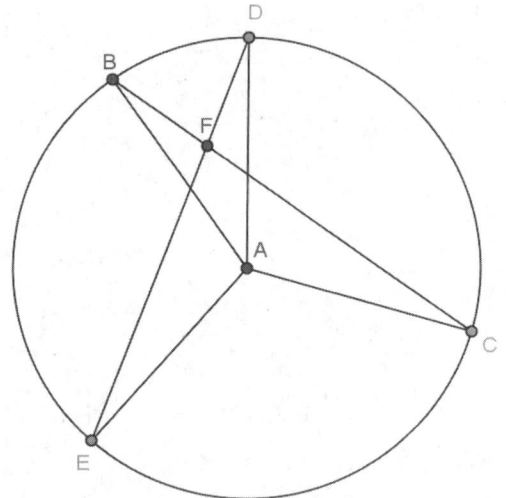

$$m\angle BED = \frac{1}{2}(m\angle\underline{\hspace{2cm}})$$

$$m\angle EBC = \frac{1}{2}(m\angle\underline{\hspace{2cm}})$$

In $\triangle EBF$,

$$m\angle BEF + m\angle EBF = m\angle\underline{\hspace{2cm}}$$

$$\frac{1}{2}(m\angle\underline{\hspace{2cm}}) + \frac{1}{2}(m\angle\underline{\hspace{2cm}}) = m\angle\underline{\hspace{2cm}}$$

$$\therefore \; m\angle DAB + m\angle EAC = 2(m\angle BFD)$$

Lesson 6: Unknown Angle Problems with Inscribed Angles in Circles

Classwork

Opening Exercise

In a circle, a chord $\overline{DE}$ and a diameter $\overline{AB}$ are extended outside of the circle to meet at point C. If $m\angle DAE = 46°$, and $m\angle DCA = 32°$, find $m\angle DEA$.

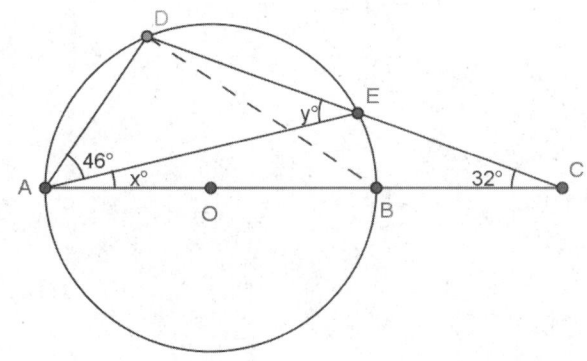

Let $m\angle DEA = y°$, $m\angle EAB = x°$

In $\triangle ABD$, $m\angle DBA =$ Reason:

$m\angle ADB =$ Reason:

$\therefore 46 + x + y + 90 =$ Reason:

$x + y =$

In $\triangle ACE$, $y = x + 32$ Reason:

$x + x + 32 =$ Reason:

$x =$

$y =$

$m\angle DEA =$

EUREKA MATH™

Exercises

Find the value x in each figure below, and describe how you arrived at the answer.

1. Hint: Thales' theorem

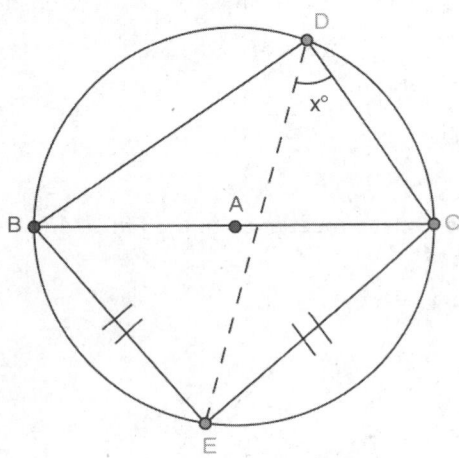

2.

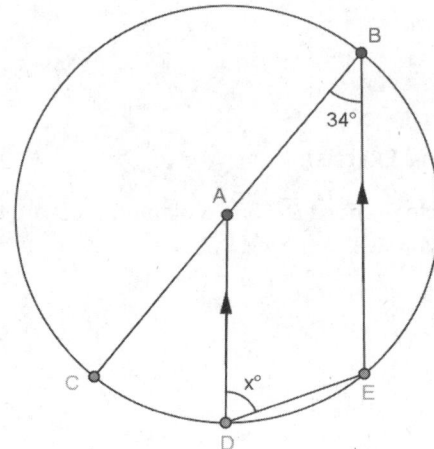

3.

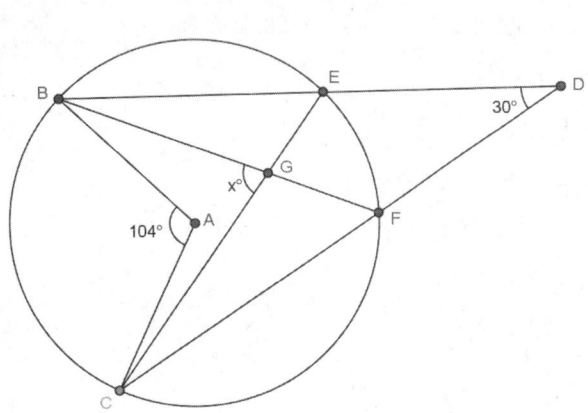

4.

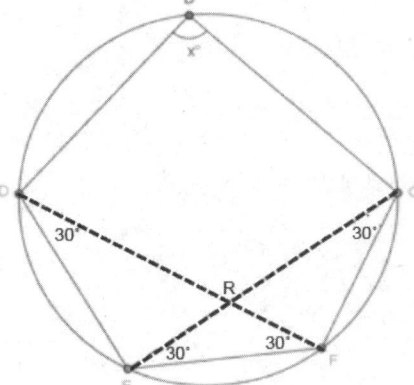

EUREKA
MATH™

> **Lesson Summary**
>
> **Theorems:**
>
> - **THE INSCRIBED ANGLE THEOREM:** The measure of a central angle is twice the measure of any inscribed angle that intercepts the same arc as the central angle.
>
> - **CONSEQUENCE OF INSCRIBED ANGLE THEOREM:** Inscribed angles that intercept the same arc are equal in measure.
>
> - If A, B, B', and C are four points with B and B' on the same side of $\overleftrightarrow{AC}$, and $\angle ABC$ and $\angle AB'C$ are congruent, then A, B, B', and C all lie on the same circle.
>
> **Relevant Vocabulary**
>
> - **CENTRAL ANGLE:** A *central angle* of a circle is an angle whose vertex is the center of a circle.
>
> - **INSCRIBED ANGLE:** An *inscribed angle* is an angle whose vertex is on a circle, and each side of the angle intersects the circle in another point.
>
> - **INTERCEPTED ARC:** An angle *intercepts an arc* if the endpoints of the arc lie on the angle, all other points of the arc are in the interior of the angle, and each side of the angle contains an endpoint of the arc. An angle inscribed in a circle intercepts exactly one arc, in particular, the arc intercepted by a right angle is the semicircle in the interior of the angle.

Problem Set

In Problems 1–5, find the value x.

1.

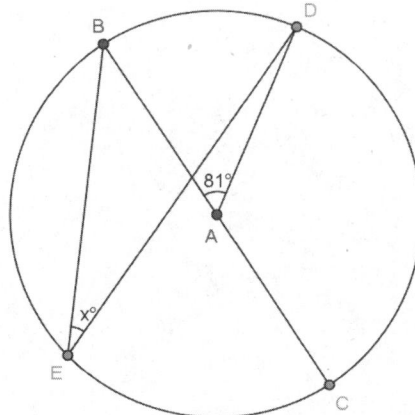

2.

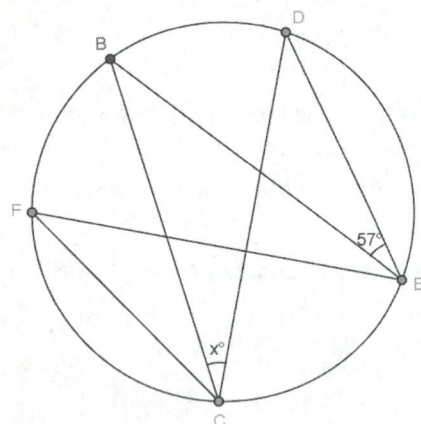

3.

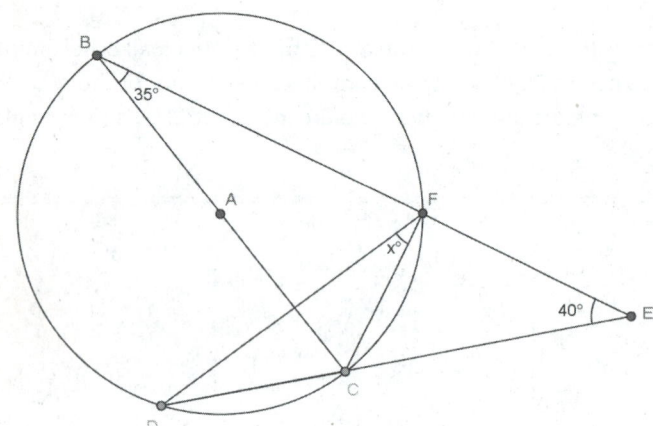

4.

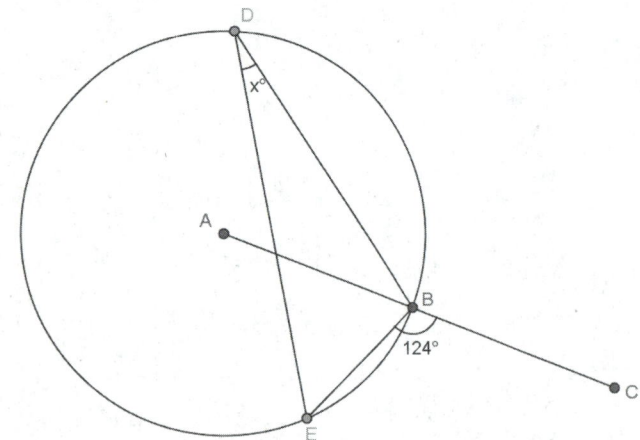

Lesson 6:   Unknown Angle Problems with Inscribed Angles in Circles

EUREKA
MATH™

5.

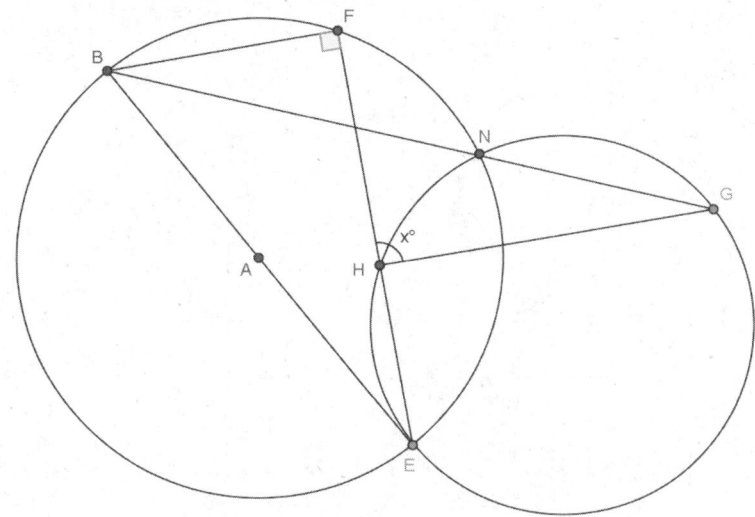

6.   If $BF = FC$, express y in terms of x.

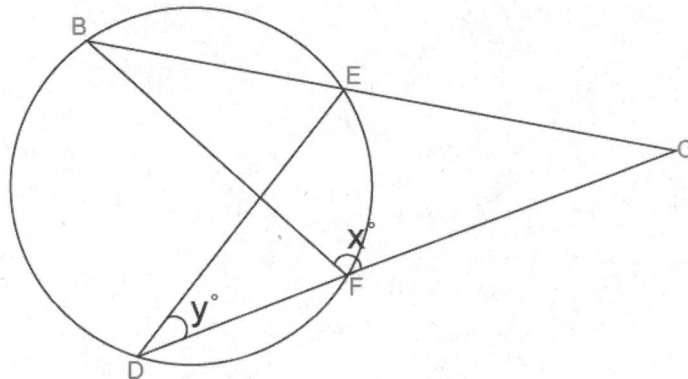

7.

    a.   Find the value of x.

    b.   Suppose the $m\angle C = a°$. Prove that $m\angle DEB = 3a°$.

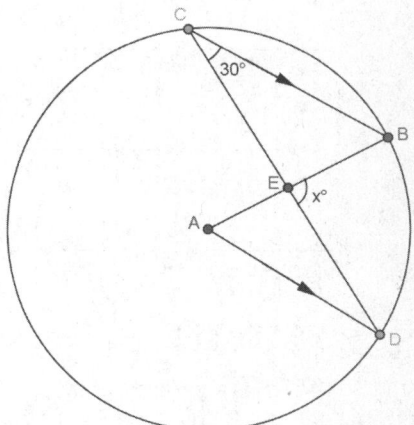

8. In the figure below, three identical circles meet at B, F, C, and E, respectively. $BF = CE$. A, B, C and F, E, D lie on straight lines.

Prove $ACDF$ is a parallelogram.

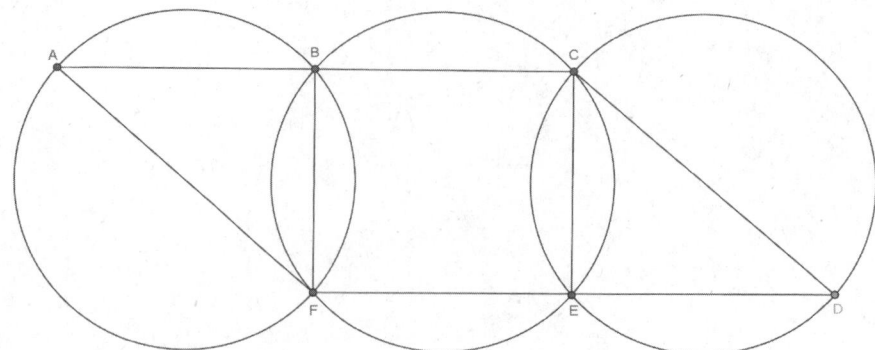

PROOF:

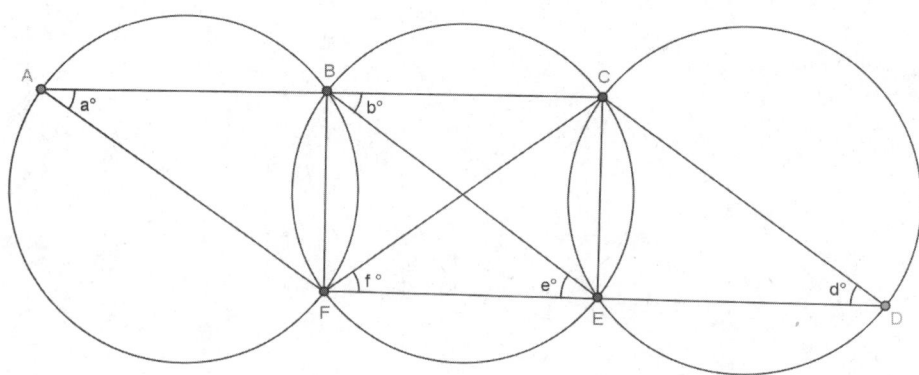

Join BE and CF.

$BF = CE$ Reason: _____

$a =$ _____ = _____ = _____ $= d$ Reason: _____

_____ = _____   Alternate interior angles are equal in measure.

$\overline{AC} \parallel \overline{FD}$

_____ = _____   Corresponding angles are equal in measure.

$\overline{AF} \parallel \overline{BE}$

_____ = _____   Corresponding angles are equal in measure.

$\overline{BE} \parallel \overline{CD}$

$\overline{AF} \parallel \overline{BE} \parallel \overline{CD}$

$ACDF$ is a parallelogram.

EUREKA
MATH™

Lesson 7: The Angle Measure of an Arc

Opening Exercise

If the measure of ∠GBF is 17°, name three other angles that have the same measure and explain why.

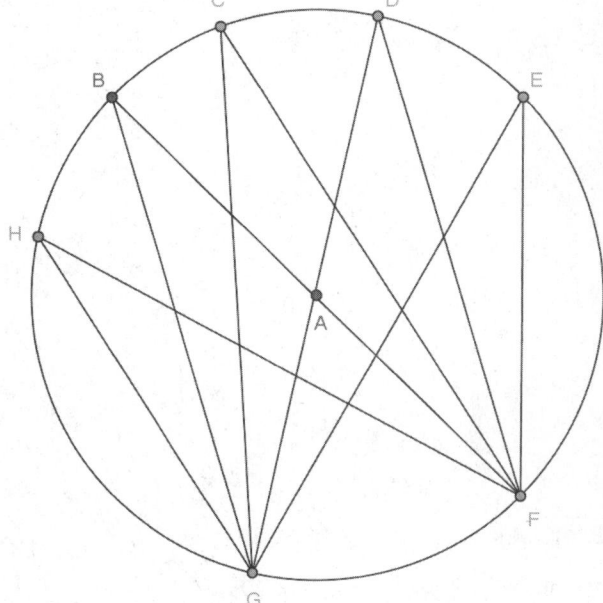

What is the measure of ∠GAF? Explain.

Can you find the measure of ∠BAD? Explain.

Example

What if we started with an angle inscribed in the minor arc between A and C?

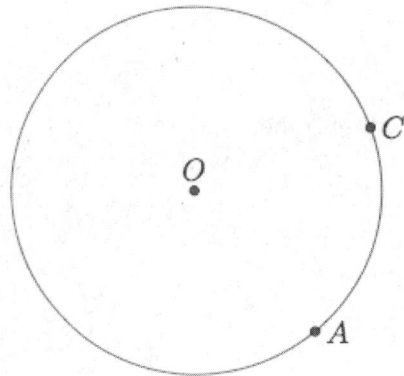

Exercises

1. In circle A, $m\widehat{BC}:m\widehat{CE}:m\widehat{ED}:m\widehat{DB} = 1:2:3:4$. Find the following angles of measure.

    a. $m\angle BAC$

    b. $m\angle DAE$

    c. $m\widehat{DB}$

    d. $m\widehat{CED}$

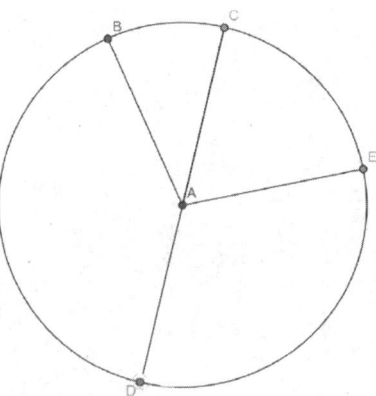

EUREKA
MATH™

© 2015 Great Minds. eureka-math.org
GEO-M3-SE-B2-1.3.0-10.2015

2. In circle B, $AB = CD$. Find the following angles of measure.

   a. $m\widehat{CD}$

   b. $m\widehat{CAD}$

   c. $m\widehat{ACD}$

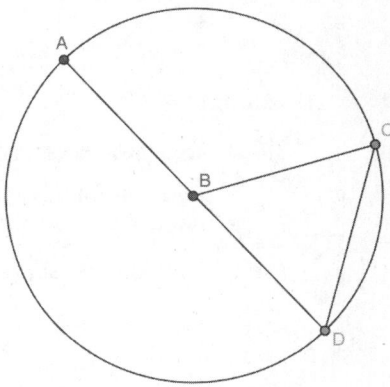

3. In circle A, $\overline{BC}$ is a diameter and $m\angle DAC = 100°$. If $m\widehat{EC} = 2m\widehat{BD}$, find the following angles of measure.

   a. $m\angle BAE$

   b. $m\widehat{EC}$

   c. $m\widehat{DEC}$

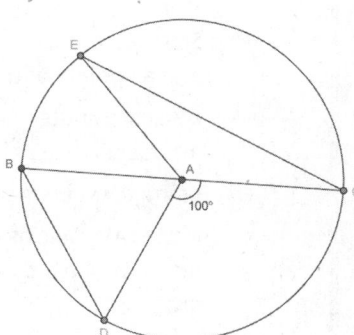

4. Given circle A with $m\angle CAD = 37°$, find the following angles of measure.

   a. $m\widehat{CBD}$

   b. $m\angle CBD$

   c. $m\angle CED$

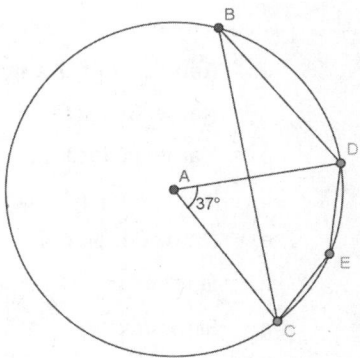

EUREKA
MATH™

Lesson 7:   The Angle Measure of an Arc

S.53

© 2015 Great Minds. eureka-math.org
GEO-M3-SE-B2-1.3.0-10.2015

Lesson Summary

Theorems:

- **INSCRIBED ANGLE THEOREM:** The measure of an inscribed angle is half the measure of its intercepted arc.

- Two arcs (of possibly different circles) are similar if they have the same angle measure. Two arcs in the same or congruent circles are congruent if they have the same angle measure.

- All circles are similar.

Relevant Vocabulary

- **ARC:** An *arc* is a portion of the circumference of a circle.

- **MINOR AND MAJOR ARC:** Let C be a circle with center O, and let A and B be different points that lie on C but are not the endpoints of the same diameter. The *minor arc* is the set containing A, B, and all points of C that are in the interior of $\angle AOB$. The *major arc* is the set containing A, B, and all points of C that lie in the exterior of $\angle AOB$.

- **SEMICIRCLE:** In a circle, let A and B be the endpoints of a diameter. A *semicircle* is the set containing A, B, and all points of the circle that lie in a given half-plane of the line determined by the diameter.

- **INSCRIBED ANGLE:** An *inscribed angle* is an angle whose vertex is on a circle and each side of the angle intersects the circle in another point.

- **CENTRAL ANGLE:** A *central angle* of a circle is an angle whose vertex is the center of a circle.

- **INTERCEPTED ARC OF AN ANGLE:** An angle *intercepts* an arc if the endpoints of the arc lie on the angle, all other points of the arc are in the interior of the angle, and each side of the angle contains an endpoint of the arc.

Problem Set

1. Given circle A with $m\angle CAD = 50°$,

    a. Name a central angle.

    b. Name an inscribed angle.

    c. Name a chord.

    d. Name a minor arc.

    e. Name a major arc.

    f. Find $m\widehat{CD}$.

    g. Find $m\widehat{CBD}$.

    h. Find $m\angle CBD$.

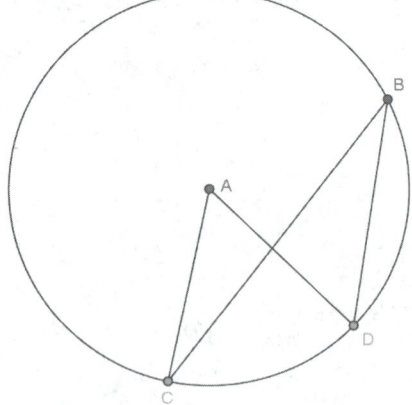

EUREKA
MATH™

2. Given circle A, find the measure of each minor arc.

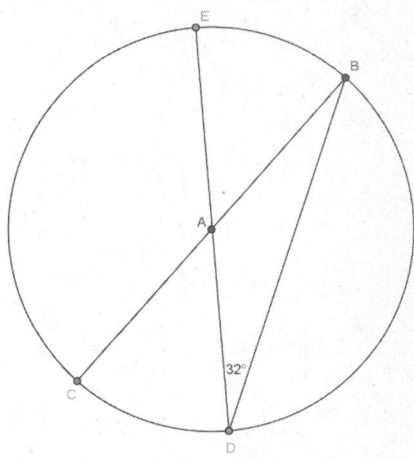

3. Given circle A, find the following measure.

    a. $m\angle BAD$
    b. $m\angle CAB$
    c. $m\widehat{BC}$
    d. $m\widehat{BD}$
    e. $m\widehat{BCD}$

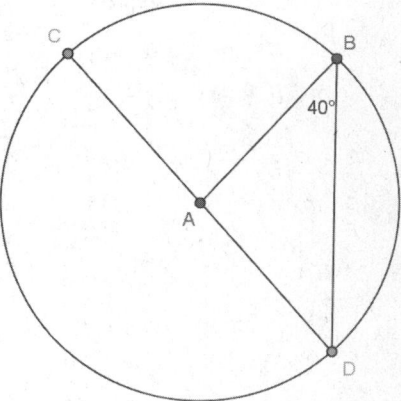

4. Find the measure of angle x.

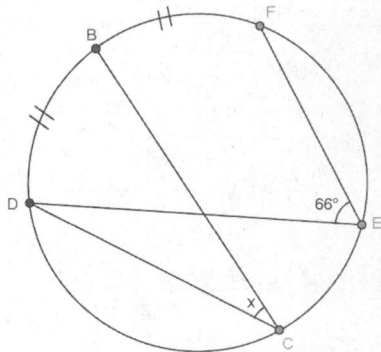

EUREKA
MATH™

5. In the figure, $m\angle BAC = 126°$ and $m\angle BED = 32°$. Find $m\angle DEC$.

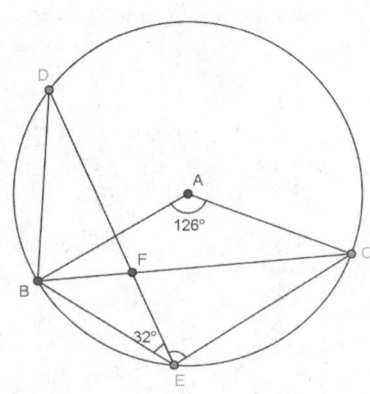

6. In the figure, $m\angle BCD = 74°$ and $m\angle BDC = 42°$. K is the midpoint of $\overset{\frown}{CB}$, and J is the midpoint of $\overset{\frown}{BD}$. Find $m\angle KBD$ and $m\angle CKJ$.

Solution: Join $BK, KC, KD, KJ, JC,$ and JD.

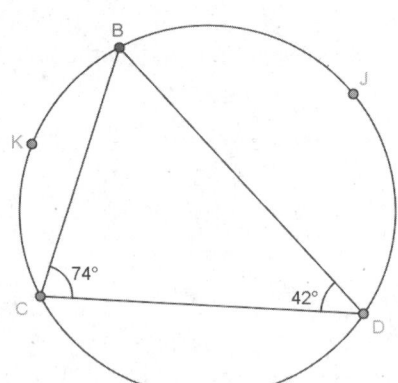

$m\overset{\frown}{BK} = m\overset{\frown}{KC}$ _____

$m\angle KDC = \dfrac{42°}{2} = 21°$ _____

$a = $ _____ _____

In $\triangle BCD, b = $ _____ _____

$c = $ _____ _____

$m\overset{\frown}{BJ} = m\overset{\frown}{JD}$ _____

$m\angle JCD = $ _____ _____

$d = $ _____ _____

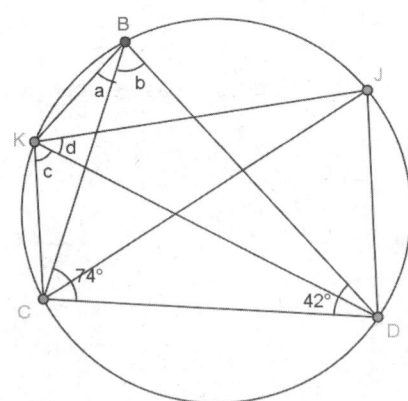

$m\angle KBD = a + b = $ _____

$m\angle CKJ = c + d = $ _____

Lesson 7:   The Angle Measure of an Arc

EUREKA
MATH™

Lesson 8: Arcs and Chords

Opening Exercise

Given circle A with $\overline{BC} \perp \overline{DE}$, $FA = 6$, and $AC = 10$. Find BF and DE. Explain your work.

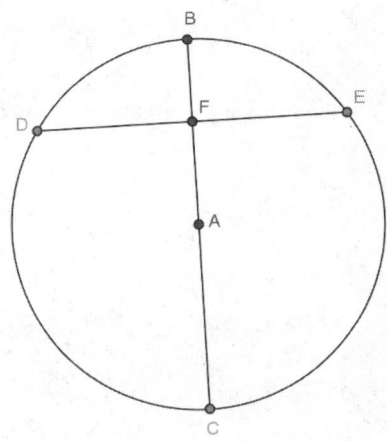

Exercises

1. Given circle A with $m\widehat{BC} = 54°$ and $\angle CDB \cong \angle DBE$, find $m\widehat{DE}$. Explain your work.

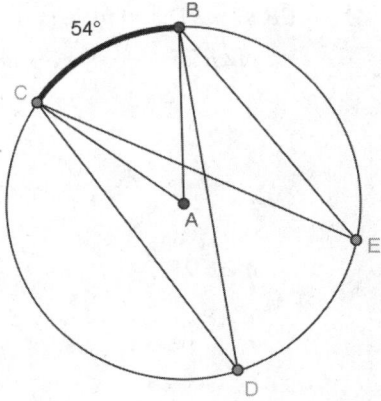

2. If two arcs in a circle have the same measure, what can you say about the quadrilateral formed by the four endpoints? Explain.

3. Find the angle measure of $\widehat{CD}$ and $\widehat{ED}$.

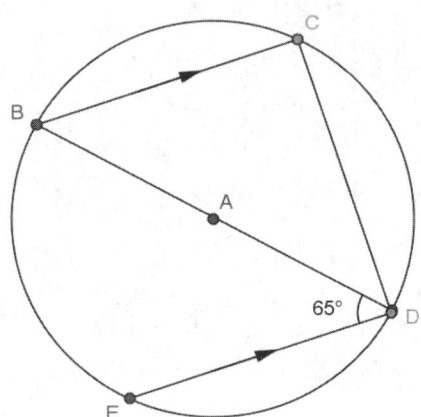

4. $m\widehat{CB} = m\widehat{ED}$ and $m\widehat{BC} : m\widehat{BD} : m\widehat{EC} = 1 : 2 : 4$. Find the following angle measures.

   a. $m\angle BCF$

   b. $m\angle EDF$

   c. $m\angle CFE$

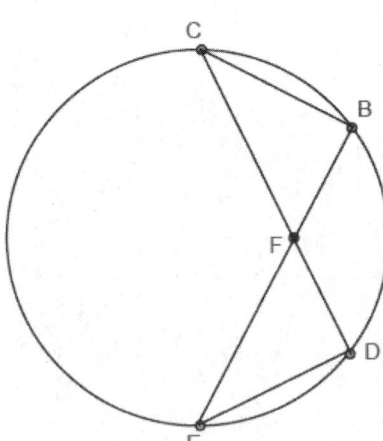

EUREKA
MATH™

5. $\overline{BC}$ is a diameter of circle A. $m\widehat{BD}:m\widehat{DE}:m\widehat{EC} = 1:3:5$. Find the following arc measures.

   a. $m\widehat{BD}$

   b. $m\widehat{DEC}$

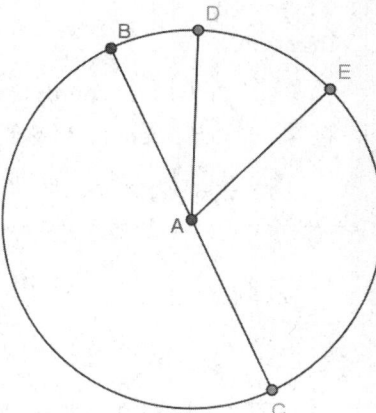

   c. $m\widehat{ECB}$

Lesson Summary

Theorems:

- Congruent chords have congruent arcs.
- Congruent arcs have congruent chords.
- Arcs between parallel chords are congruent.

Problem Set

1. Find the following arc measures.

   a. $m\widehat{CE}$

   b. $m\widehat{BD}$

   c. $m\widehat{ED}$

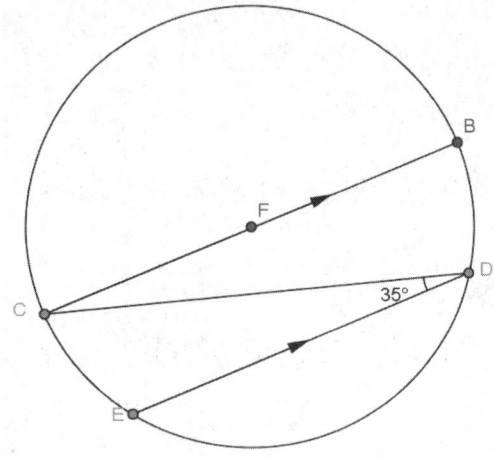

2. In circle A, $\overline{BC}$ is a diameter, $m\widehat{CE} = m\widehat{ED}$, and $m\angle CAE = 32°$.

   a. Find $m\angle CAD$.

   b. Find $m\angle ADC$.

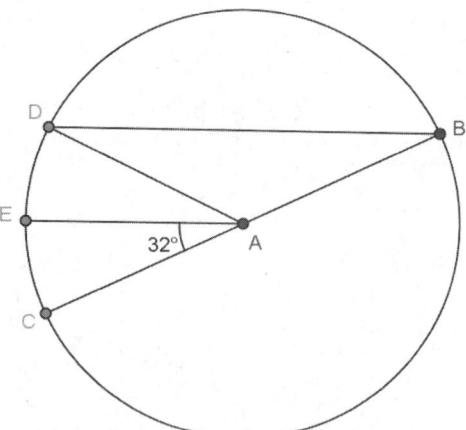

EUREKA
MATH™

© 2015 Great Minds. eureka-math.org
GEO-M3-SE-B2-1.3.0-10.2015

3. In circle A, $\overline{BC}$ is a diameter, $2m\widehat{CE} = m\widehat{ED}$, and $\overline{BC} \parallel \overline{DE}$. Find $m\angle CDE$.

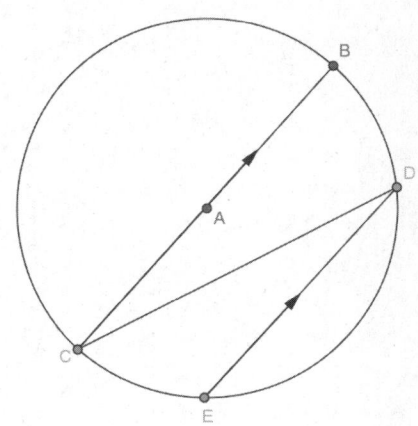

4. In circle A, $\overline{BC}$ is a diameter and $m\widehat{CE} = 68°$.

    a. Find $m\widehat{CD}$.

    b. Find $m\angle DBE$.

    c. Find $m\angle DCE$

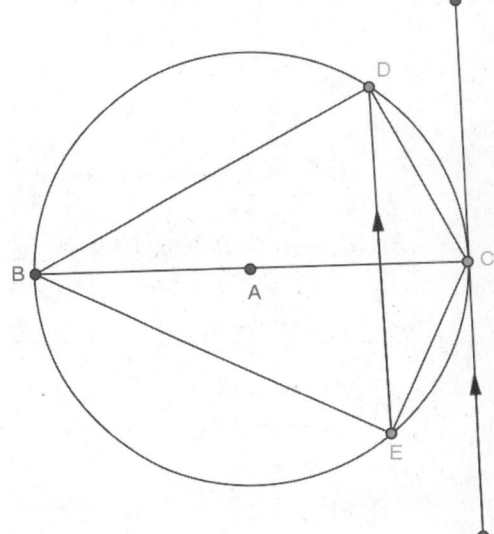

5. In the circle given, $\widehat{BC} \cong \widehat{ED}$. Prove $\overline{BE} \cong \overline{DC}$.

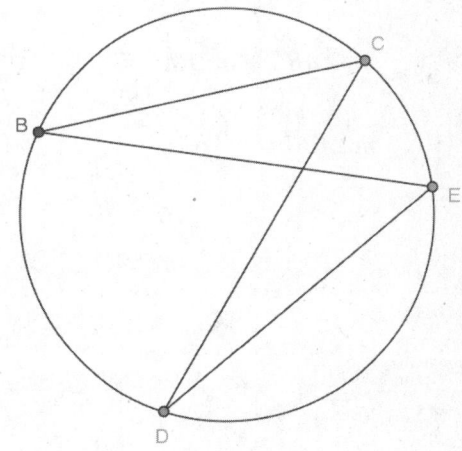

6. Given circle A with $\overline{AD} \parallel \overline{CE}$, show $\overset{\frown}{BD} \cong \overset{\frown}{DE}$.

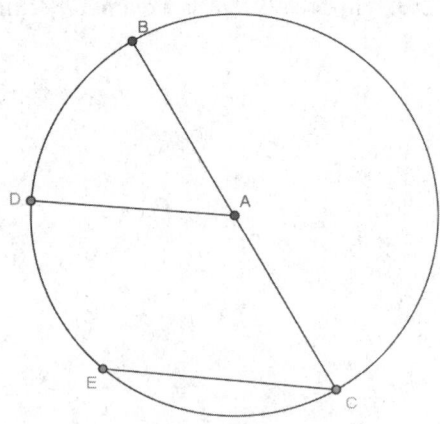

7. In circle A, $\overline{AB}$ is a radius, $\overset{\frown}{BC} \cong \overset{\frown}{BD}$, and $m\angle CAD = 54°$. Find $m\angle ABC$. Complete the proof.

$BC = BD$ _____

$m\angle \underline{\quad} = m\angle \underline{\quad}$ _____

$m\angle BAC + m\angle CAD + m\angle BAD = \underline{\quad}$

$2m\angle \underline{\quad} + 54° = 360°$ _____

$m\angle BAC = \underline{\quad}$

$AB = AC$ _____

$m\angle \underline{\quad} = m\angle \underline{\quad}$ _____

$2m\angle ABC + m\angle BAC = \underline{\quad}$ _____

$m\angle ABC = \underline{\quad}$

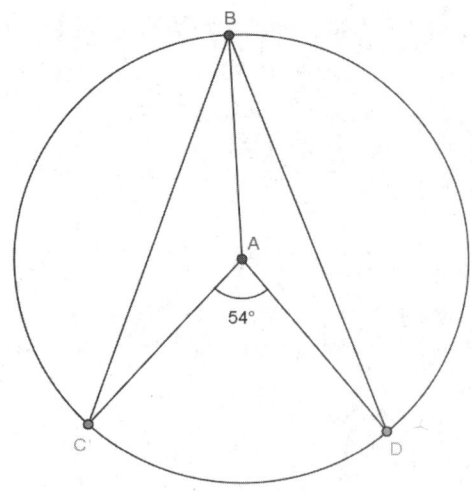

Lesson 9: Arc Length and Areas of Sectors

Classwork

Example 1

a. What is the length of the arc of degree that measures $60°$ in a circle of radius 10 cm?

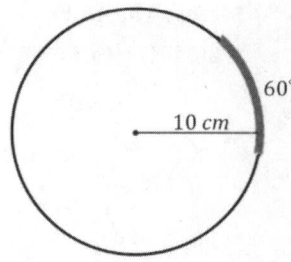

b. Given the concentric circles with center A and with $m\angle A = 60°$, calculate the arc length intercepted by $\angle A$ on each circle. The inner circle has a radius of 10, and each circle has a radius 10 units greater than the previous circle.

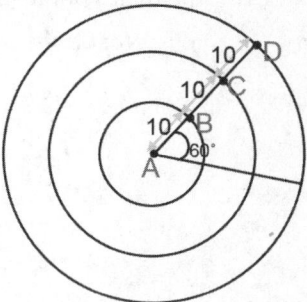

c. An arc, again of degree measure $60°$, has an arc length of 5π cm. What is the radius of the circle on which the arc sits?

d. Give a general formula for the length of an arc of degree measure $x°$ on a circle of radius r.

EUREKA MATH™

e.   Is the length of an arc intercepted by an angle proportional to the radius? Explain.

Sector: Let $\widehat{AB}$ be an arc of a circle with center O and radius r. The union of all segments $\overline{OP}$, where P is any point of $\widehat{AB}$, is called a *sector*.

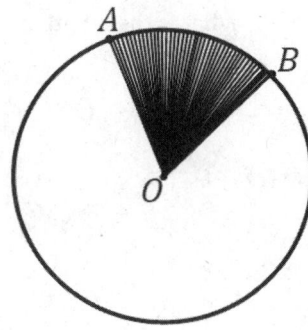

Exercise 1

1.   The radius of the following circle is 36 cm, and the $m\angle ABC = 60°$.

a.   What is the arc length of $\widehat{AC}$?

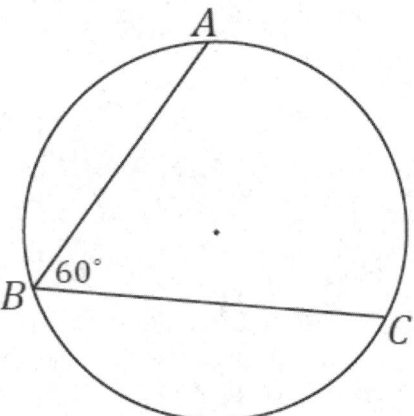

b.   What is the radian measure of the central angle?

Lesson 9:   Arc Length and Areas of Sectors

EUREKA
MATH

Example 2

    a.   Circle O has a radius of 10 cm. What is the area of the circle? Write the formula.

    b.   What is the area of half of the circle? Write and explain the formula.

    c.   What is the area of a quarter of the circle? Write and explain the formula.

    d.   Make a conjecture about how to determine the area of a sector defined by an arc measuring 60°.

    e.   Circle O has a minor arc $\widehat{AB}$ with an angle measure of 60°. Sector AOB has an area of 24π. What is the radius of circle O?

    f.   Give a general formula for the area of a sector defined by an arc of angle measure $x°$ on a circle of radius r.

© 2015 Great Minds. eureka-math.org
GEO-M3-SE-B2-1.3.0-10.2015

Exercises 2–3

2. The area of sector AOB in the following image is 28π cm^2. Find the measurement of the central angle labeled $x°$.

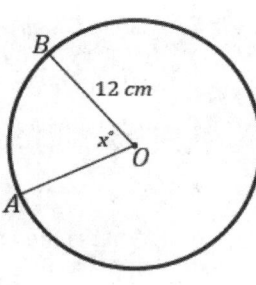

3. In the following figure of circle O, $m\angle AOC = 108°$ and $\widehat{AB} = \widehat{AC} = 10$ cm.

   a. Find $m\angle OAB$.

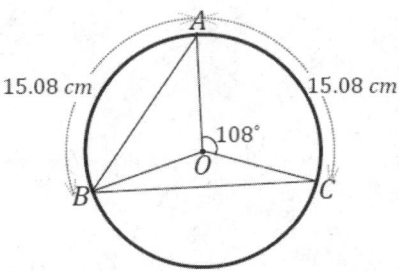

   b. Find $m\widehat{BC}$.

   c. Find the area of sector BOC.

EUREKA
MATH™

Lesson Summary

Relevant Vocabulary

- **ARC:** An *arc* is any of the following three figures—a minor arc, a major arc, or a semicircle.

- **LENGTH OF AN ARC:** The *length of an arc* is the circular distance around the arc.

- **MINOR AND MAJOR ARC:** In a circle with center O, let A and B be different points that lie on the circle but are not the endpoints of a diameter. The *minor arc* between A and B is the set containing A, B, and all points of the circle that are in the interior of $\angle AOB$. The *major arc* is the set containing A, B, and all points of the circle that lie in the exterior of $\angle AOB$.

- **RADIAN:** A *radian* is the measure of the central angle of a sector of a circle with arc length of one radius length.

- **SECTOR:** Let $\widehat{AB}$ be an arc of a circle with center O and radius r. The union of the segments $\overline{OP}$, where P is any point on $\widehat{AB}$, is called a sector. $\widehat{AB}$ is called the arc of the sector, and r is called its radius.

- **SEMICIRCLE:** In a circle, let A and B be the endpoints of a diameter. A *semicircle* is the set containing A, B, and all points of the circle that lie in a given half-plane of the line determined by the diameter.

Problem Set

1. P and Q are points on the circle of radius 5 cm, and the measure of arc $\widehat{PQ}$ is 72°. Find, to one decimal place, each of the following.

   a. The length of $\widehat{PQ}$

   b. The ratio of the arc length to the radius of the circle

   c. The length of chord $\overline{PQ}$

   d. The distance of the chord $\overline{PQ}$ from the center of the circle

   e. The perimeter of sector POQ

   f. The area of the wedge between the chord $\overline{PQ}$ and $\widehat{PQ}$

   g. The perimeter of this wedge

2. What is the radius of a circle if the length of a 45° arc is 9π?

3. $\widehat{AB}$ and $\widehat{CD}$ both have an angle measure of 30°, but their arc lengths are not the same. $OB = 4$ and $BD = 2$.

   a. What are the arc lengths of $\widehat{AB}$ and $\widehat{CD}$?

   b. What is the ratio of the arc length to the radius for both of these arcs? Explain.

   c. What are the areas of the sectors AOB and COD?

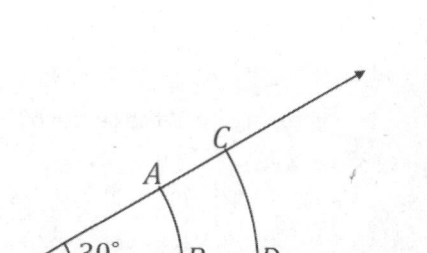

4. In the circles shown, find the value of x. Figures are not drawn to scale.

a. The circles have central angles of equal measure.

b.

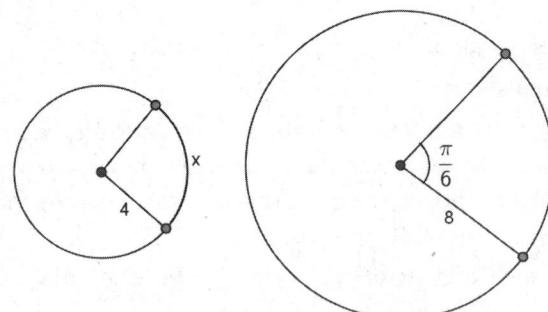

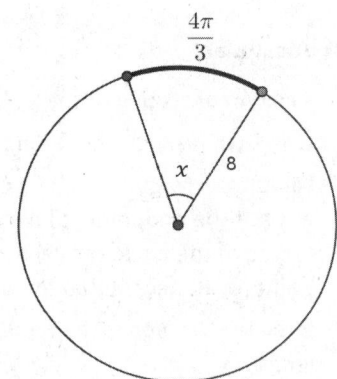

c.

d.

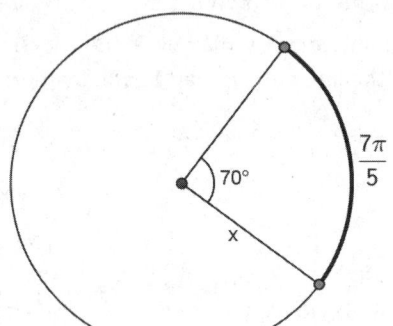

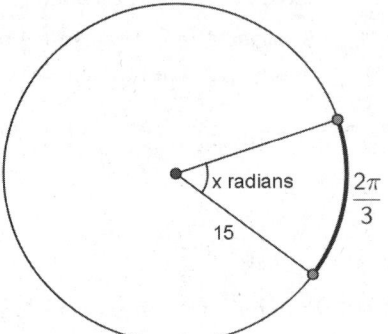

5. The concentric circles all have center A. The measure of the central angle is $45°$. The arc lengths are given.

a. Find the radius of each circle.

b. Determine the ratio of the arc length to the radius of each circle, and interpret its meaning.

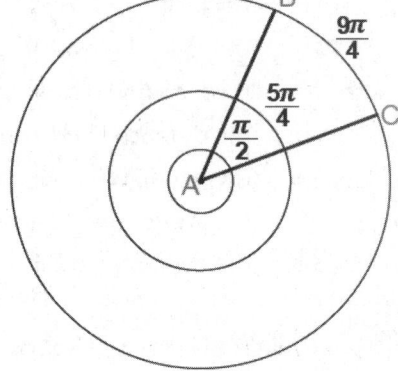

6. In the figure, if the length of $\overset{\frown}{PQ}$ is 10 cm, find the length of $\overset{\frown}{QR}$.

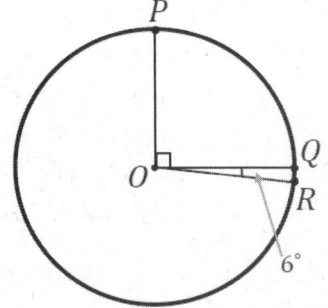

EUREKA
MATH™

7. Find, to one decimal place, the areas of the shaded regions.

    a.

    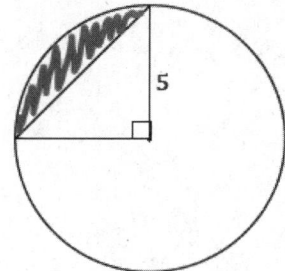

    b. The following circle has a radius of 2.

    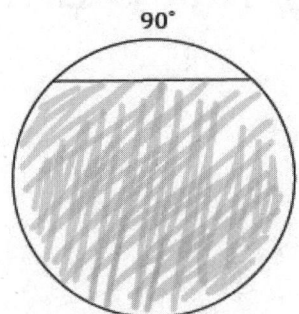

    c.

    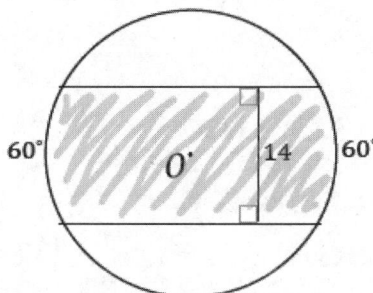

This page intentionally left blank

Lesson 10: Unknown Length and Area Problems

Opening Exercise

In the following figure, a cylinder is carved out from within another cylinder of the same height; the bases of both cylinders share the same center.

    a.   Sketch a cross section of the figure parallel to the base.

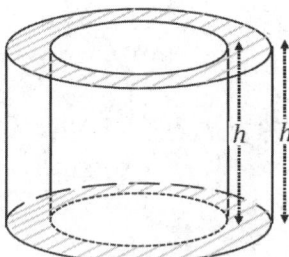

    b.   Mark and label the shorter of the two radii as r and the longer of the two radii as s.

        Show how to calculate the area of the shaded region, and explain the parts of the expression.

Exercises

1.   Find the area of the following annulus.

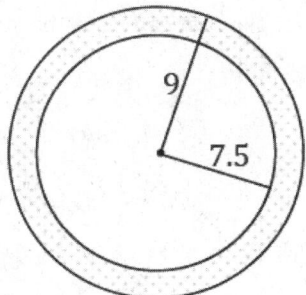

2. The larger circle of an annulus has a diameter of 10 cm, and the smaller circle has a diameter of 7.6 cm. What is the area of the annulus?

3. In the following annulus, the radius of the larger circle is twice the radius of the smaller circle. If the area of the following annulus is 12π units^2, what is the radius of the larger circle?

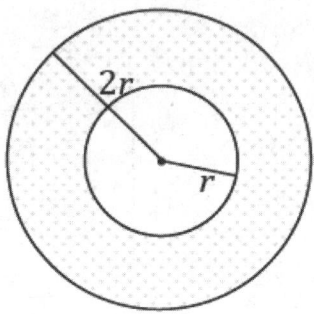

4. An ice cream shop wants to design a super straw to serve with its extra thick milkshakes that is double both the width and thickness of a standard straw. A standard straw is 4 mm in diameter and 0.5 mm thick.

   a. What is the cross-sectional (parallel to the base) area of the new straw (round to the nearest hundredth)?

   b. If the new straw is 10 cm long, what is the maximum volume of milkshake that can be in the straw at one time (round to the nearest hundredth)?

c. A large milkshake is 32 fl. oz. (approximately 950 mL). If Corbin withdraws the full capacity of a straw 10 times a minute, what is the minimum amount of time that it will take him to drink the milkshake (round to the nearest minute)?

5. In the circle given, $\overline{ED}$ is the diameter and is perpendicular to chord $\overline{CB}$. $DF = 8$ cm, and $FE = 2$ cm. Find AC, BC, $m\angle CAB$, the arc length of $\overparen{CEB}$, and the area of sector CAB (round to the nearest hundredth, if necessary).

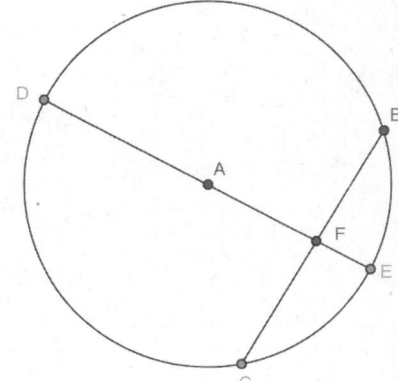

6. Given circle A with $\angle BAC \cong \angle BAD$, find the following (round to the nearest hundredth, if necessary).
    a. $m\overparen{CD}$

    b. $m\overparen{CBD}$

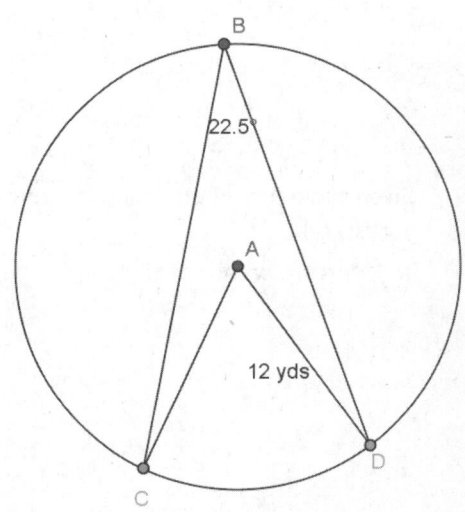

EUREKA MATH™

c. $m\overset{\frown}{BCD}$

d. Arc length $\overset{\frown}{CD}$

e. Arc length $\overset{\frown}{CBD}$

f. Arc length $\overset{\frown}{BCD}$

g. Area of sector CAD

7. Given circle A, find the following (round to the nearest hundredth, if necessary).

   a. Circumference of circle A

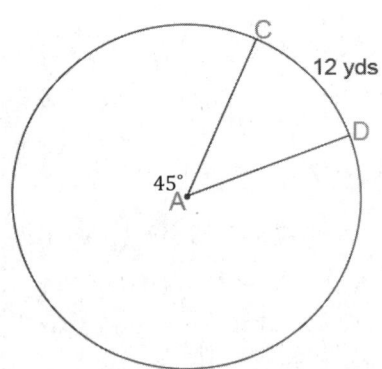

Lesson 10: Unknown Length and Area Problems

EUREKA
MATH™

b.   Radius of circle A

c.   Area of sector CAD

8.   Given circle A, find the following (round to the nearest hundredth, if necessary).

a. $m\angle CAD$

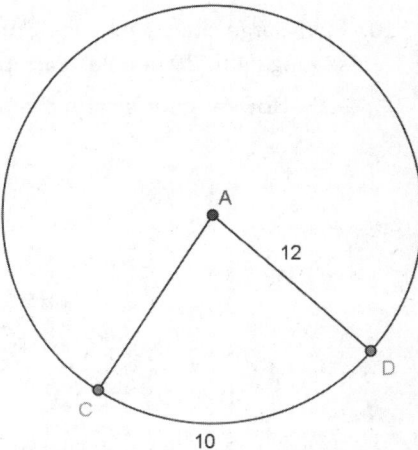

b.   Area of sector CD

9. Find the area of the shaded region (round to the nearest hundredth).

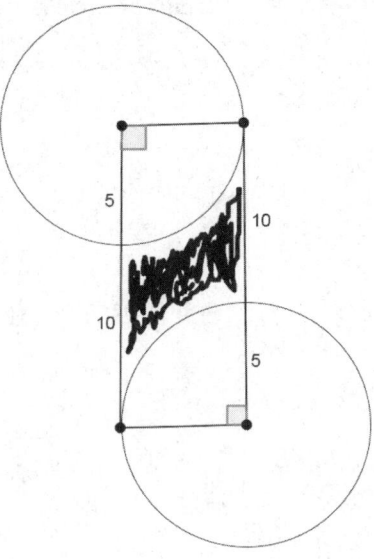

10. Many large cities are building or have built mega Ferris wheels. One is 600 feet in diameter and has 48 cars each seating up to 20 people. Each time the Ferris wheel turns θ degrees, a car is in a position to load.

    a. How far does a car move with each rotation of θ degrees (round to the nearest whole number)?

    b. What is the value of θ in degrees?

Lesson 10:   Unknown Length and Area Problems

© 2015 Great Minds. eureka-math.org
GEO-M3-SE-B2-1.3.0-10.2015

11. △ ABC is an equilateral triangle with edge length 20 cm. D, E, and F are midpoints of the sides. The vertices of the triangle are the centers of the circles creating the arcs shown. Find the following (round to the nearest hundredth).

a.   Area of the sector with center A

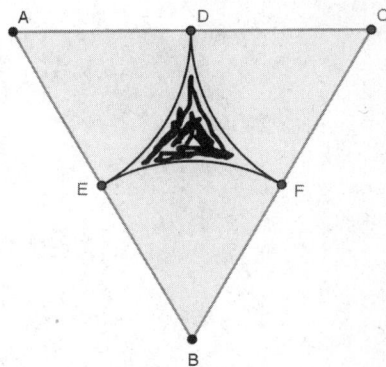

b.   Area of △ ABC

c.   Area of the shaded region

d.   Perimeter of the shaded region

12. In the figure shown, $AC = BF = 5$ cm, $GH = 2$ cm, and $m\angle HAI = 30°$. Find the area inside the rectangle but outside of the circles (round to the nearest hundredth).

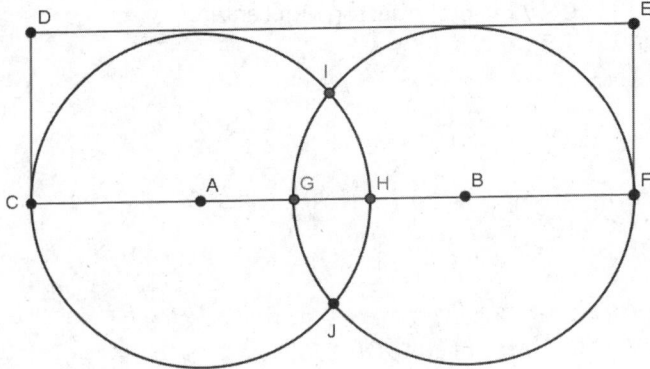

13. This is a picture of a piece of a mosaic tile. If the radius of each smaller circle is 1 inch, find the area of the red section, the white section, and the blue section (round to the nearest hundredth).

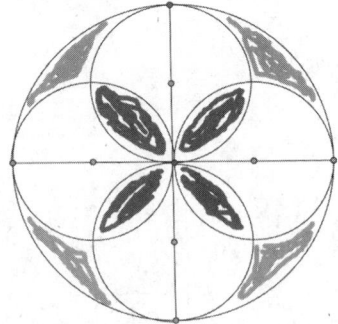

EUREKA
MATH™

Problem Set

1. Find the area of the shaded region if the diameter is 32 inches (round to the nearest hundredth).

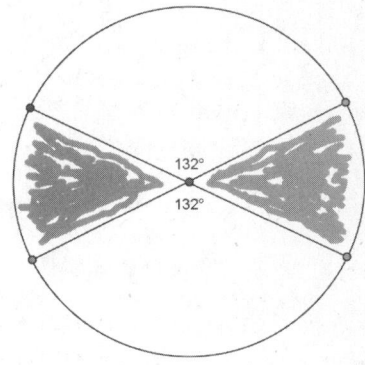

2. Find the area of the entire circle given the area of the sector.

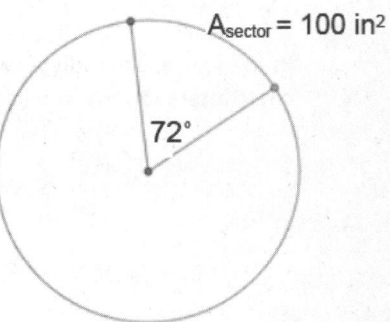

3. $\overset{\frown}{DF}$ and $\overset{\frown}{BG}$ are arcs of concentric circles with $\overline{BD}$ and $\overline{FG}$ lying on the radii of the larger circle. Find the area of the region (round to the nearest hundredth).

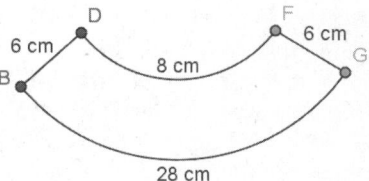

4. Find the radius of the circle, as well as x, y, and z (leave angle measures in radians and arc length in terms of pi). Note that C and D do *not* lie on a diameter.

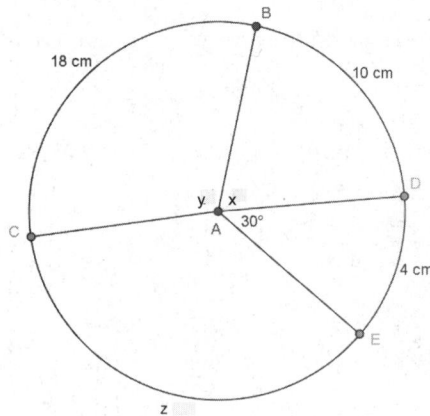

5. In the figure, the radii of two concentric circles are 24 cm and 12 cm. $m\widehat{DAE} = 120°$. If a chord $\overline{DE}$ of the larger circle intersects the smaller circle only at C, find the area of the shaded region in terms of π.

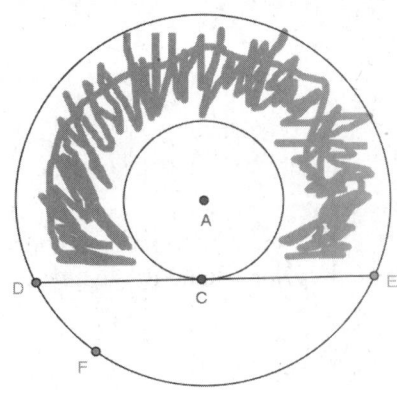

© 2015 Great Minds. eureka-math.org
GEO-M3-SE-B2-1.3.0-10.2015

EUREKA
MATH™

Lesson 11: Properties of Tangents

Exercises

1. $\overline{CD}$ and $\overline{CE}$ are tangent to circle A at points D and E, respectively. Use a two-column proof to prove $a = b$.

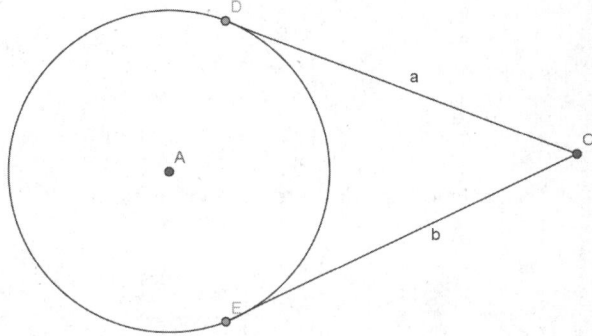

© 2015 Great Minds. eureka-math.org
GEO-M3-SE-B2-1.3.0-10.2015

2. In circle A, the radius is 9 mm and $BC = 12$ mm.

    a. Find AC.

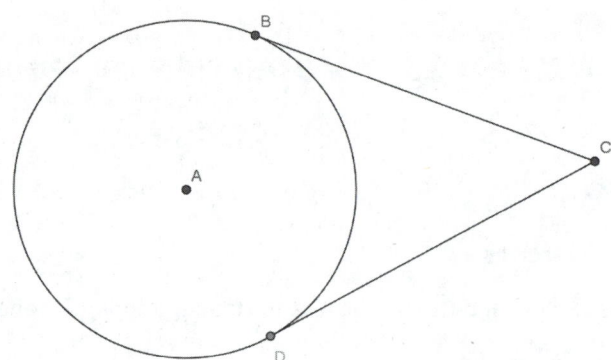

    b. Find the area of $\triangle ACD$.

    c. Find the perimeter of quadrilateral $ABCD$.

3. In circle A, $EF = 12$ and $AE = 13$. $AE:AC = 1:3$.

    a. Find the radius of the circle.

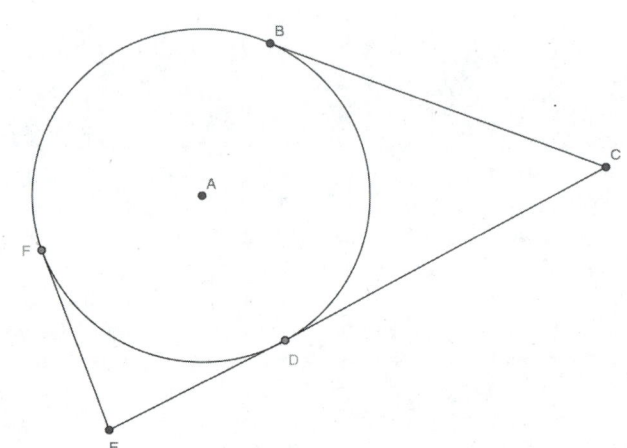

    b. Find BC (round to the nearest whole number).

    c. Find EC.

© 2015 Great Minds. eureka-math.org
GEO-M3-SE-B2-1.3.0-10.2015

Lesson Summary

THEOREMS:

- A tangent line to a circle is perpendicular to the radius of the circle drawn to the point of tangency.

- A line through a point on a circle is tangent at the point if, and only if, it is perpendicular to the radius drawn to the point of tangency.

Relevant Vocabulary

- **INTERIOR OF A CIRCLE:** The *interior of a circle* with center O *and* radius r is the set of all points in the plane whose distance from the point O is less than r.

  A point in the interior of a circle is said to be inside the circle. A disk is the union of the circle with its interior.

- **EXTERIOR OF A CIRCLE:** The *exterior of a circle* with center O and radius r is the set of all points in the plane whose distance from the point O is greater than r.

  A point exterior to a circle is said to be outside the circle.

- **TANGENT TO A CIRCLE:** A *tangent line to a circle* is a line in the same plane that intersects the circle in one and only one point. This point is called the point of tangency.

- **TANGENT SEGMENT/RAY:** A segment is a *tangent segment* to a circle if the line that contains it is tangent to the circle and one of the end points of the segment is a point of tangency. A ray is called a *tangent ray* to a circle if the line that contains it is tangent to the circle and the vertex of the ray is the point of tangency.

- **SECANT TO A CIRCLE:** A secant line to a circle is a line that intersects a circle in exactly two points.

- **POLYGON INSCRIBED IN A CIRCLE:** A polygon is inscribed in a circle if all of the vertices of the polygon lie on the circle.

- **CIRCLE INSCRIBED IN A POLYGON:** A circle is inscribed in a polygon if each side of the polygon is tangent to the circle.

Problem Set

1. If $AB = 5$, $BC = 12$, and $AC = 13$, is $\overleftrightarrow{BC}$ tangent to circle A at point B? Explain.

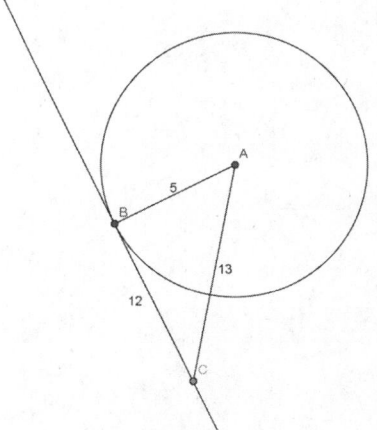

2. $\overleftrightarrow{BC}$ is tangent to circle A at point B. $DC = 9$ and $BC = 15$.

    a.   Find the radius of the circle.

    b.   Find AC.

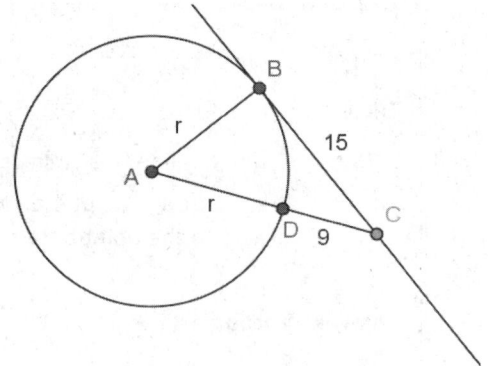

3. A circular pond is fenced on two opposite sides ($\overline{CD}$, $\overline{FE}$) with wood and the other two sides with metal fencing. If all four sides of fencing are tangent to the pond, is there more wood or metal fencing used?

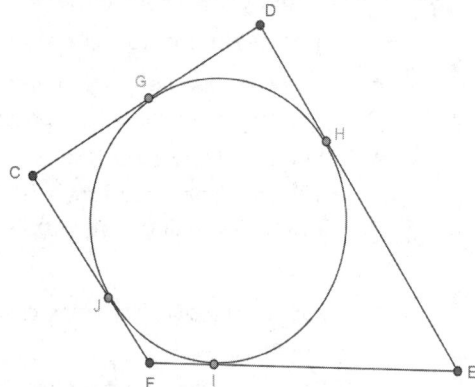

4. Find x if the line shown is tangent to the circle at point B.

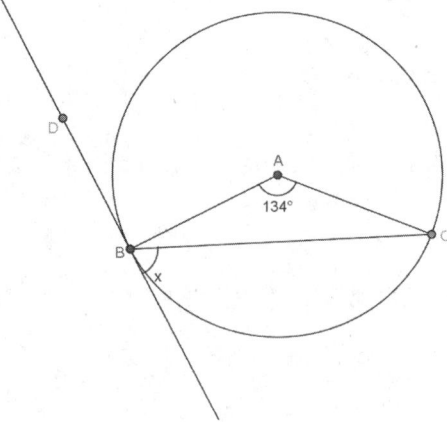

EUREKA
MATH™

5. $\overleftrightarrow{PC}$ is tangent to the circle at point C, and $CD = DE$.

   a. Find x ($m\widehat{CD}$).

   b. Find y ($m\angle CFE$).

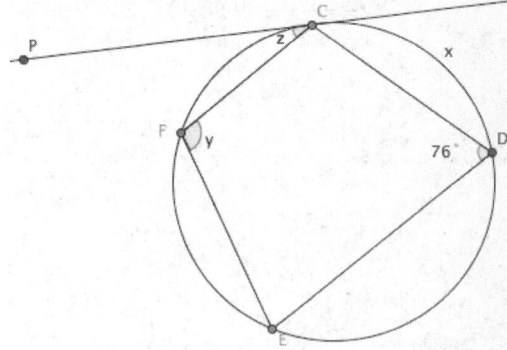

6. Construct two lines tangent to circle A through point B.

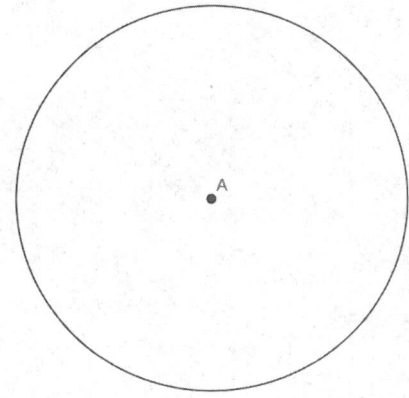

7. Find x, the length of the common tangent line between the two circles (round to the nearest hundredth).

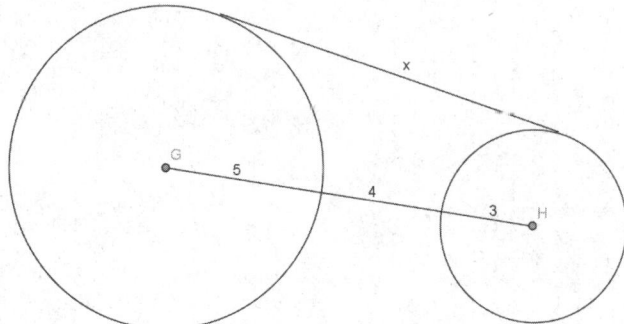

EUREKA
MATH™

Lesson 11:   Properties of Tangents

S.85

© 2015 Great Minds. eureka-math.org
GEO-M3-SE-B2-1.3.0-10.2015

8. $\overline{EF}$ is tangent to both circles A and C. The radius of circle A is 9, and the radius of circle C is 5. The circles are 2 units apart. Find the length of $\overline{EF}$, or x (round to the nearest hundredth).

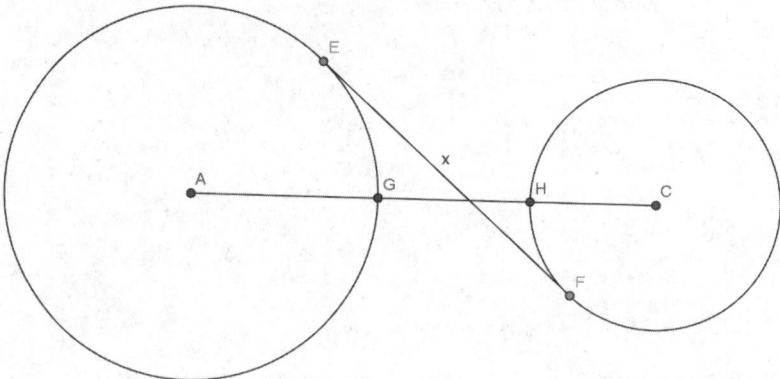

EUREKA
MATH™

Lesson 12: Tangent Segments

Opening Exercise

In the diagram, what do you think the length of z could be? How do you know?

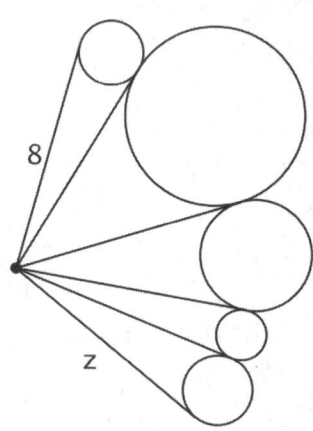

Example

In each diagram, try to draw a circle with center D that is tangent to both rays of $\angle BAC$.

a.

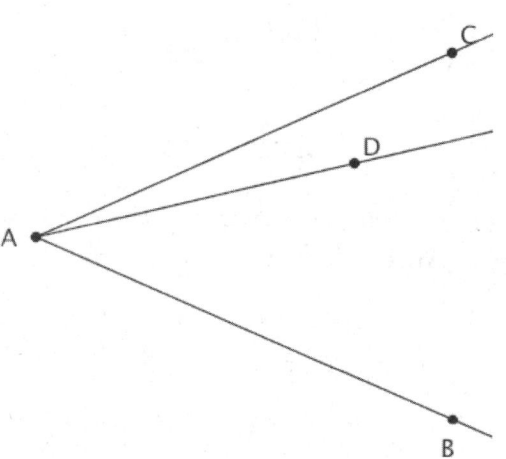

b.

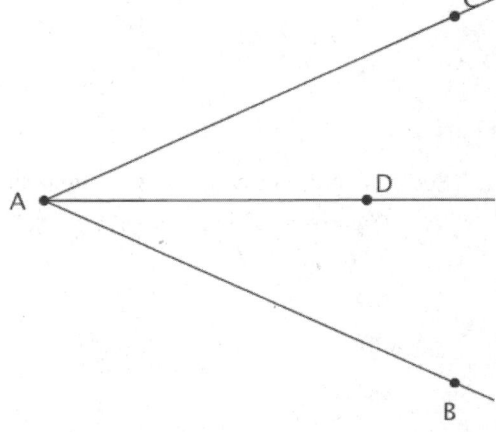

c.

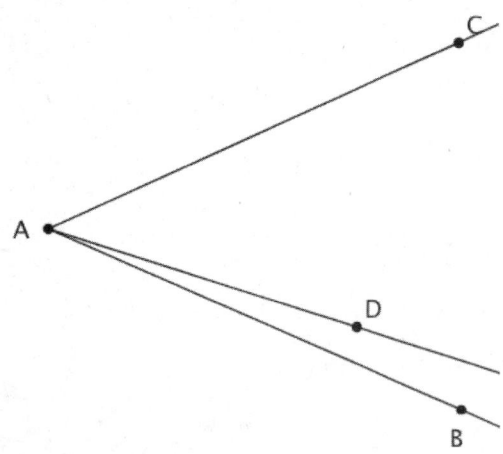

In which diagrams did it seem impossible to draw such a circle? Why did it seem impossible?

What do you conjecture about circles tangent to both rays of an angle? Why do you think that?

Exercises

1. You conjectured that *if a circle is tangent to both rays of a circle, then the center lies on the angle bisector.*

    a. Rewrite this conjecture in terms of the notation suggested
        by the diagram.

        Given:

        Need to show:

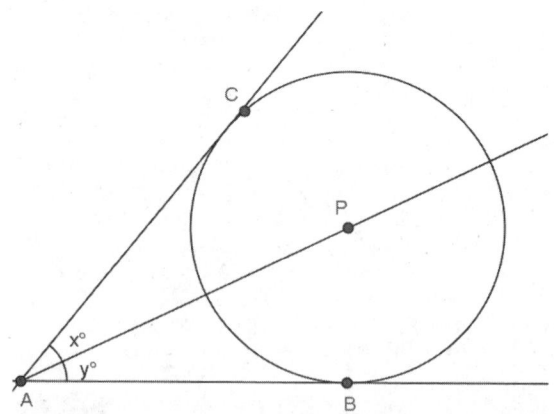

EUREKA
MATH™

b.   Prove your conjecture using a two-column proof.

2. An angle is shown below.

a.   Draw at least three different circles that are tangent to both rays of the given angle.

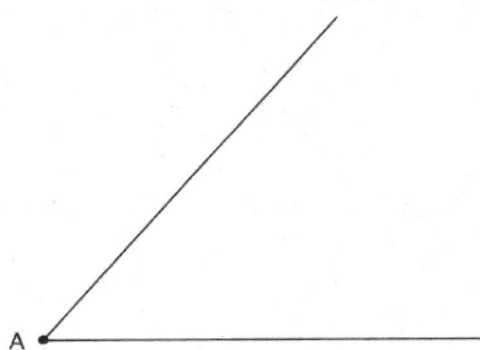

b.   Label the center of one of your circles with P. How does the distance between P and the rays of the angle compare to the radius of the circle? How do you know?

3. Construct as many circles as you can that are tangent to both the given angles at the same time. You can extend the rays as needed. These two angles share a side.

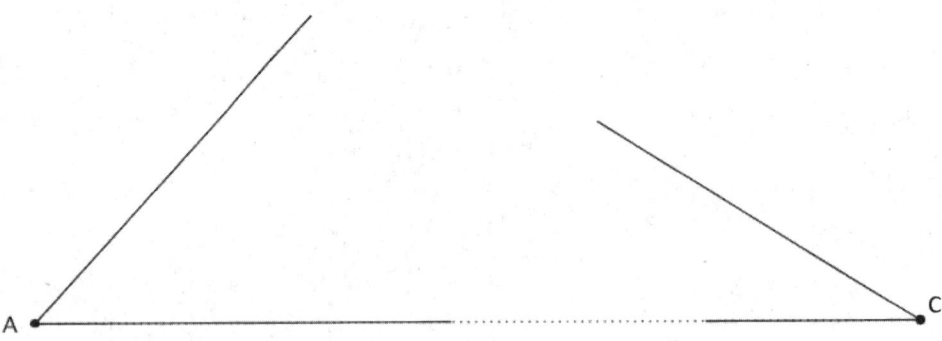

Explain how many circles you can draw to meet the above conditions and how you know.

4. In a triangle, let P be the location where two angle bisectors meet. Must P be on the third angle bisector as well? Explain your reasoning.

Lesson 12:   Tangent Segments

EUREKA
MATH™

5. Using a straightedge, draw a large triangle ABC.

    a. Construct a circle inscribed in the given triangle.

    b. Explain why your construction works.

    c. Do you know another name for the intersection of the angle bisectors in relation to the triangle?

> **Lesson Summary**
>
> **THEOREMS:**
>
> - The two tangent segments to a circle from an exterior point are congruent.
> - If a circle is tangent to both rays of an angle, then its center lies on the angle bisector.
> - Every triangle contains an inscribed circle whose center is the intersection of the triangle's angle bisectors.

Problem Set

1. On a piece of paper, draw a circle with center A and a point, C, outside of the circle.

   a. How many tangents can you draw from C to the circle?

   b. Draw two tangents from C to the circle, and label the tangency points D and E. Fold your paper along the line AC. What do you notice about the lengths of $\overline{CD}$ and $\overline{CE}$? About the measures of $\angle DCA$ and $\angle ECA$?

   c. $\overline{AC}$ is the _____ of $\angle DCE$.

   d. $\overline{CD}$ and $\overline{CE}$ are tangent to circle A. Find AC.

2. In the figure, the three segments are tangent to the circle at points B, F, and G. If $y = \frac{2}{3}x$, find x, y, and z.

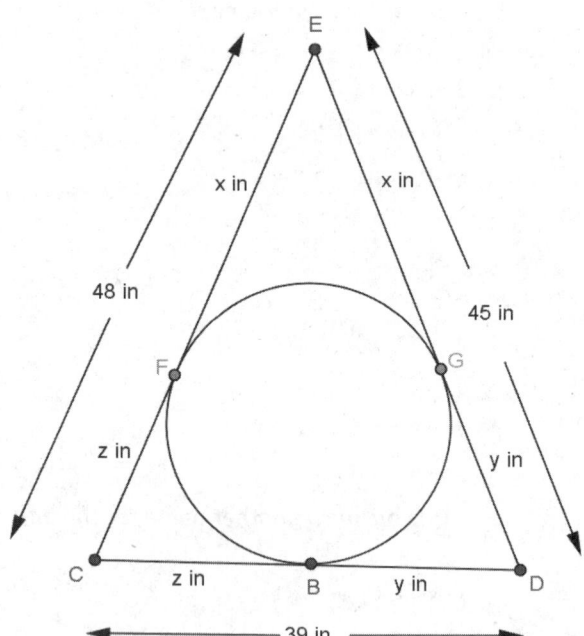

EUREKA
MATH™

3. In the figure given, the three segments are tangent to the circle at points J, I, and H.

a. Prove $GF = GJ + HF$.

b. Find the perimeter of $\triangle\,GCF$.

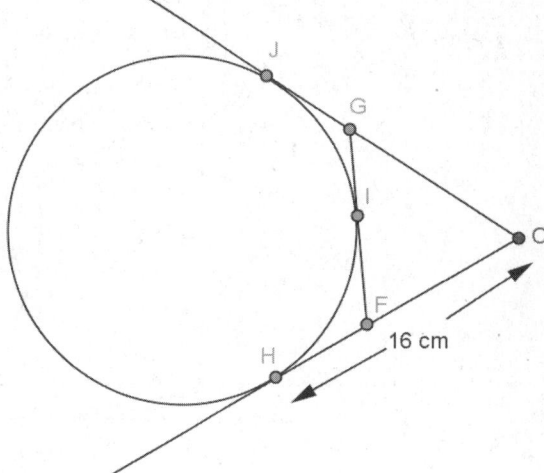

4. In the figure given, the three segments are tangent to the circle at points F, B, and G. Find DE.

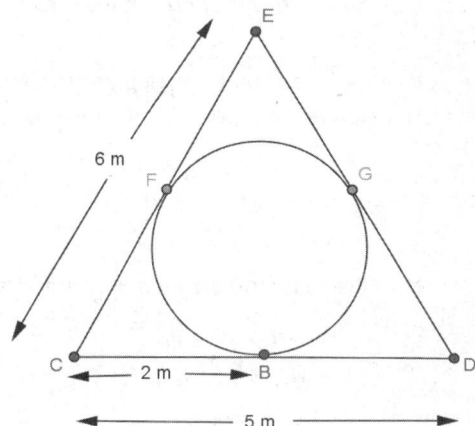

5. $\overleftrightarrow{EF}$ is tangent to circle A. If points C and D are the intersection points of circle A and any line parallel to $\overleftrightarrow{EF}$, answer the following.

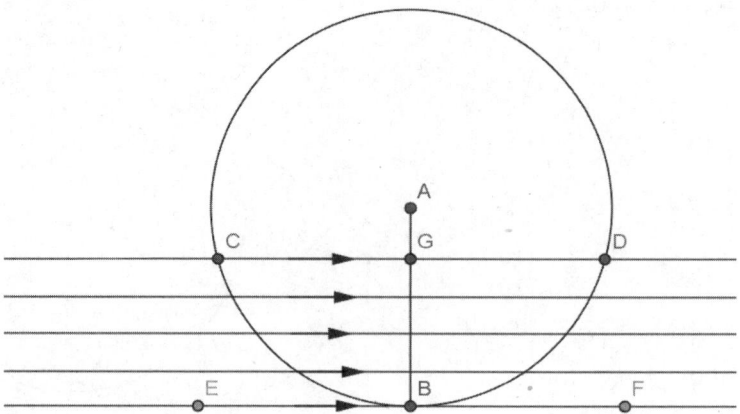

a. Does $CG = GD$ for any line parallel to $\overleftrightarrow{EF}$? Explain.

b. Suppose that $\overleftrightarrow{CD}$ coincides with $\overleftrightarrow{EF}$. Would C, G, and D all coincide with B?

c. Suppose C, G, and D have now reached B, so $\overleftrightarrow{CD}$ is tangent to the circle. What is the angle between $\overleftrightarrow{CD}$ and $\overline{AB}$?

d. Draw another line tangent to the circle from some point, P, in the exterior of the circle. If the point of tangency is point T, what is the measure of $\angle PTA$?

6. The segments are tangent to circle A at points B and D. $\overline{ED}$ is a diameter of the circle.

a. Prove $\overline{BE} \parallel \overline{CA}$.

b. Prove quadrilateral $ABCD$ is a kite.

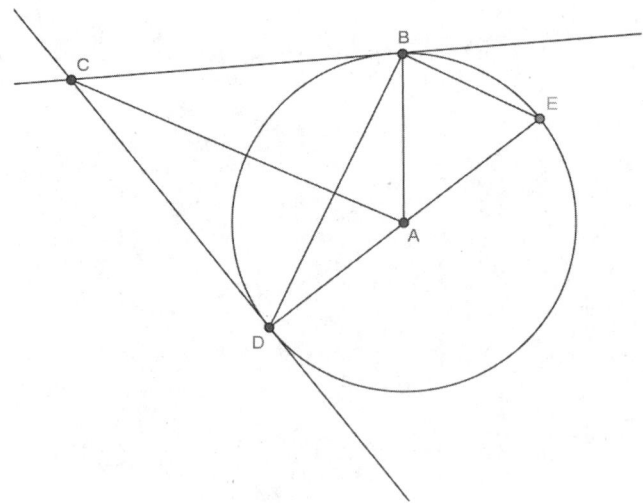

EUREKA
MATH™

7. In the diagram shown, $\overleftrightarrow{BH}$ is tangent to the circle at point B. What is the relationship between $\angle DBH$, the angle between the tangent and a chord, and the arc subtended by that chord and its inscribed angle $\angle DCB$?

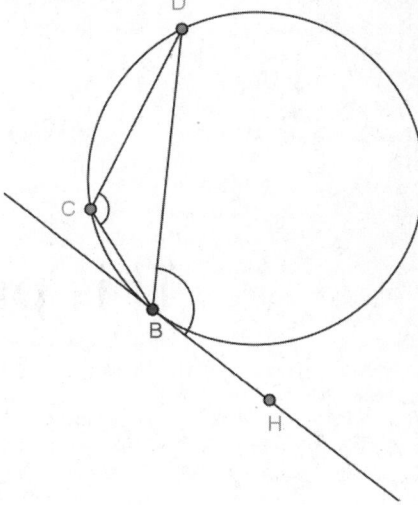

This page intentionally left blank

Lesson 13: The Inscribed Angle Alternate—A Tangent Angle

Opening Exercise

In circle A, $m\widehat{BD} = 56°$, and $\overline{BC}$ is a diameter. Find the listed measure, and explain your answer.

a. $m\angle BDC$

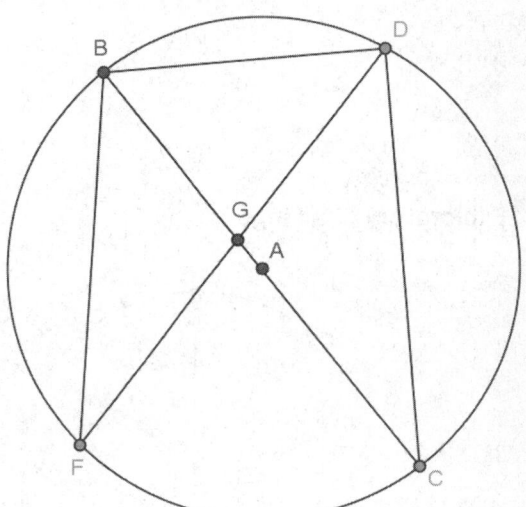

b. $m\angle BCD$

c. $m\angle DBC$

d. $m\angle BFG$

e. $m\widehat{BC}$

f. $m\widehat{DC}$

EUREKA
MATH™

Lesson 13:   The Inscribed Angle Alternate—A Tangent Angle

S.97

© 2015 Great Minds. eureka-math.org
GEO-M3-SE-B2-1.3.0-10.2015

g. Does ∠BGD measure 56°? Explain.

h. How do you think we could determine the measure of ∠BGD?

Exploratory Challenge

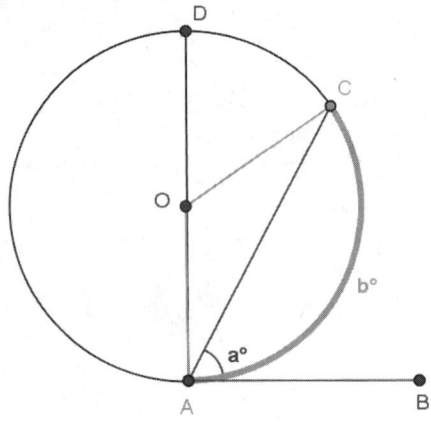

Diagram 1

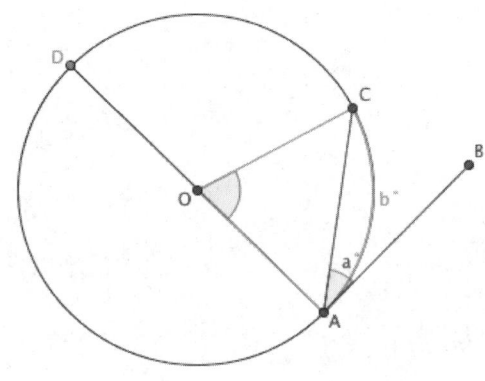

Diagram 2

Examine the diagrams shown. Develop a conjecture about the relationship between a and b.

Test your conjecture by using a protractor to measure a and b.

	a	b
Diagram 1		
Diagram 2		

EUREKA
MATH

Do your measurements confirm the relationship you found in your homework?

If needed, revise your conjecture about the relationship between a and b:

Now, test your conjecture further using the circle below.

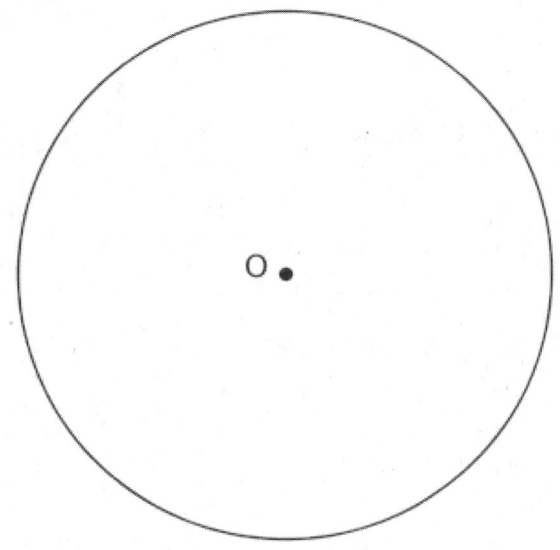

a	b

Now, we will prove your conjecture, which is stated below as a theorem.

THE TANGENT-SECANT THEOREM: Let A be a point on a circle, let $\overrightarrow{AB}$ be a tangent ray to the circle, and let C be a point on the circle such that $\overleftrightarrow{AC}$ is a secant to the circle. If $a = m\angle BAC$ and b is the angle measure of the arc intercepted by $\angle BAC$, then $a = \frac{1}{2}b$.

Given circle O with tangent $\overleftrightarrow{AB}$, prove what we have just discovered using what you know about the properties of a circle and tangent and secant lines.

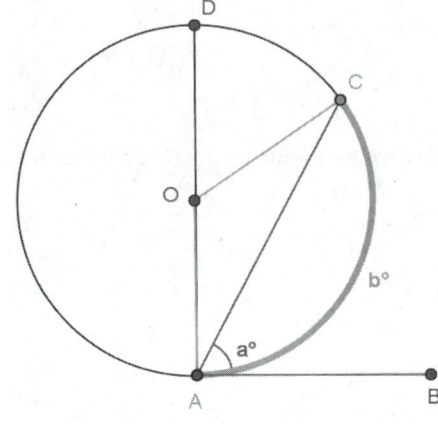

a.   Draw triangle AOC. What is the measure of $\angle AOC$? Explain.

b.   What is the measure of $\angle OAB$? Explain.

c.   Express the measure of the remaining two angles of triangle AOC in terms of a and explain.

d.   What is the measure of $\angle AOC$ in terms of a? Show how you got the answer.

e.   Explain to your neighbor what we have just proven.

EUREKA
MATH™

Exercises

Find x, y, a, b, and/or c.

1.

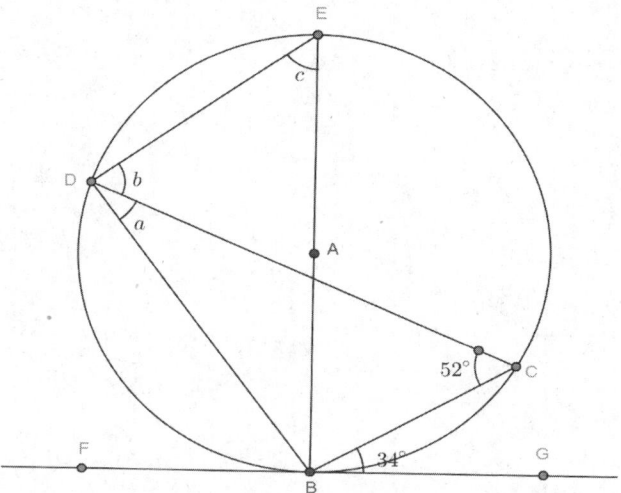

2.

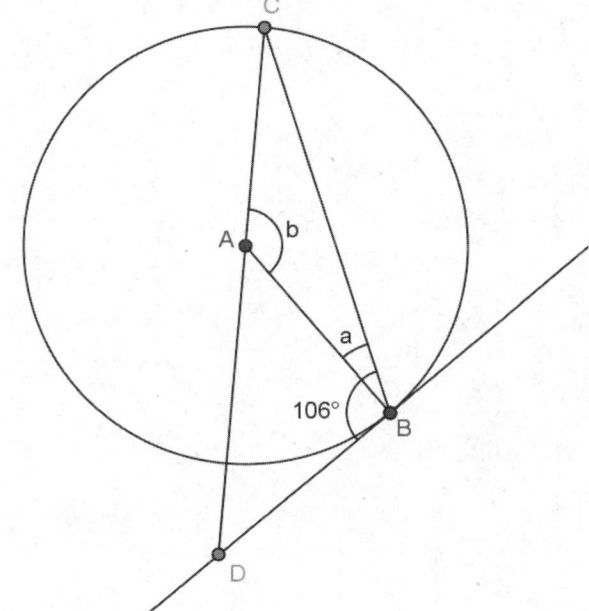

EUREKA
MATH™

© 2015 Great Minds. eureka-math.org
GEO-M3-SE-B2-1.3.0-10.2015

3.

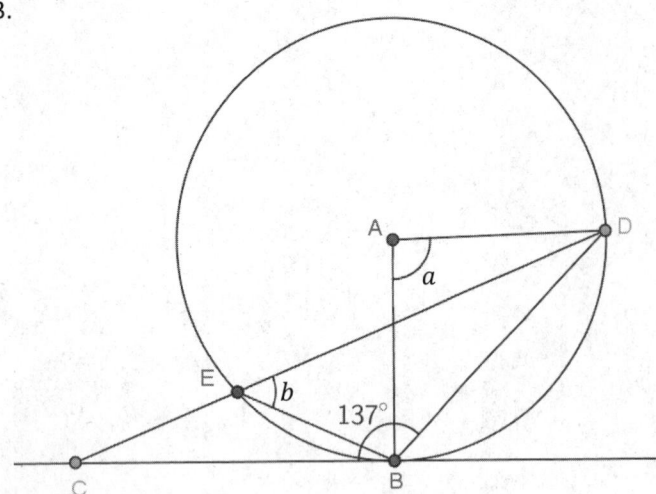

4.

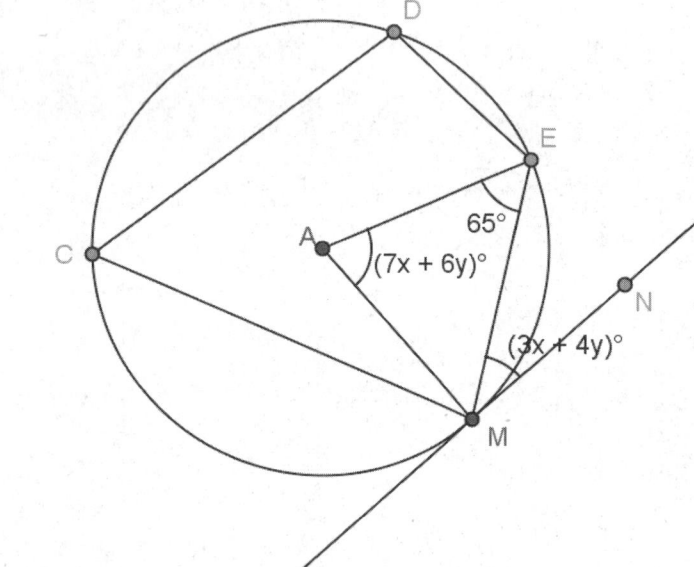

Lesson 13:   The Inscribed Angle Alternate—A Tangent Angle

EUREKA
MATH™

5.

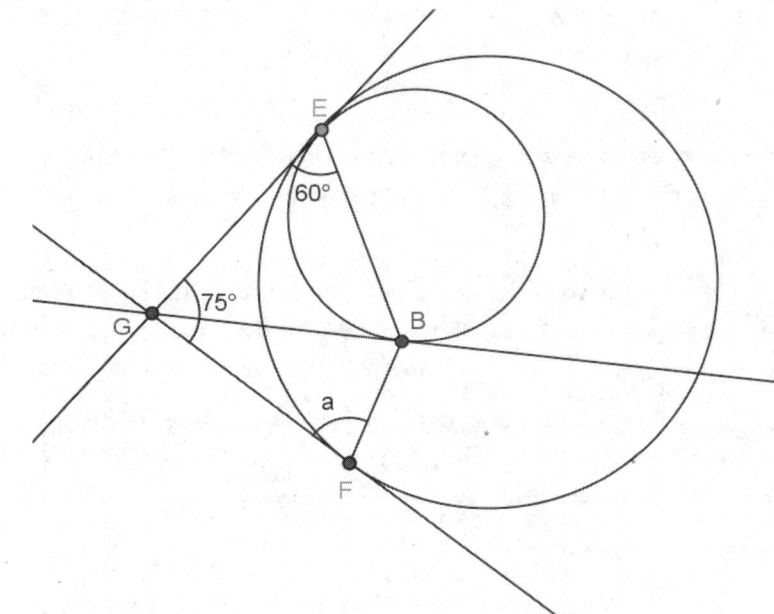

Problem Set

In Problems 1–9, solve for a, b, and/or c.

1.

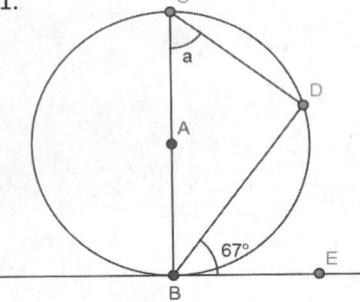

2.

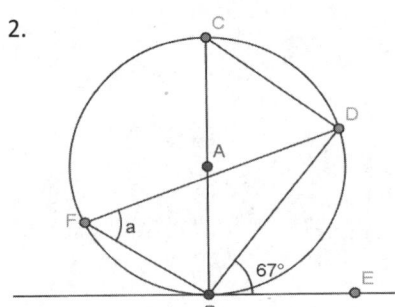

3.

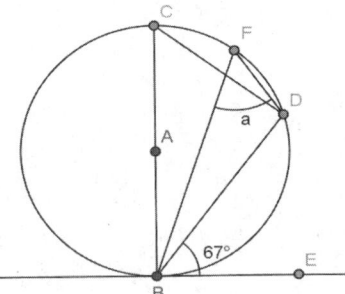

4.

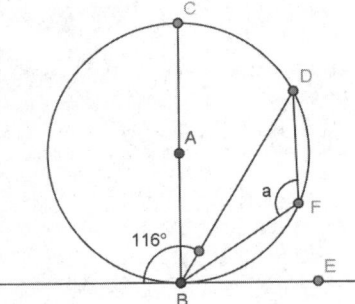

5.

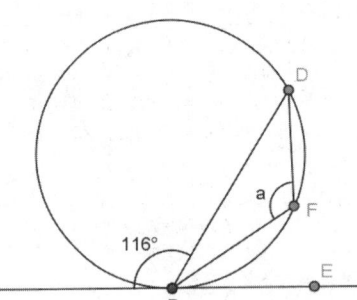

6.

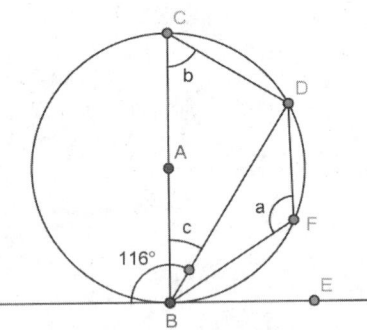

EUREKA
MATH™

7.

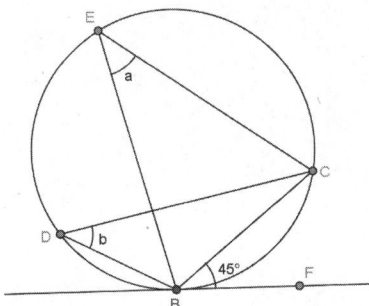

8.

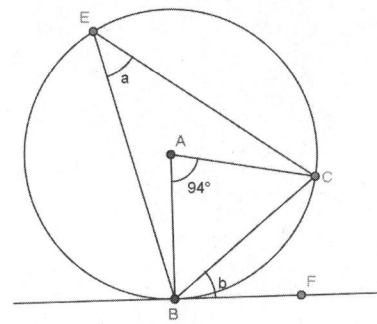

9.

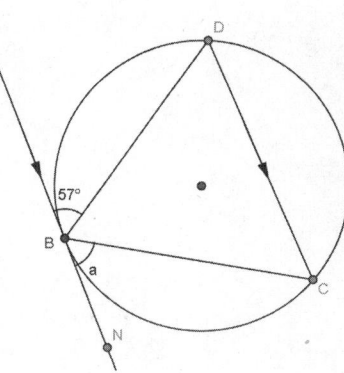

10. $\overleftrightarrow{BH}$ is tangent to circle A. $\overline{DF}$ is a diameter. Find the angle measurements.

   a. $m\angle BCD$

   b. $m\angle BAF$

   c. $m\angle BDA$

   d. $m\angle FBH$

   e. $m\angle BGF$

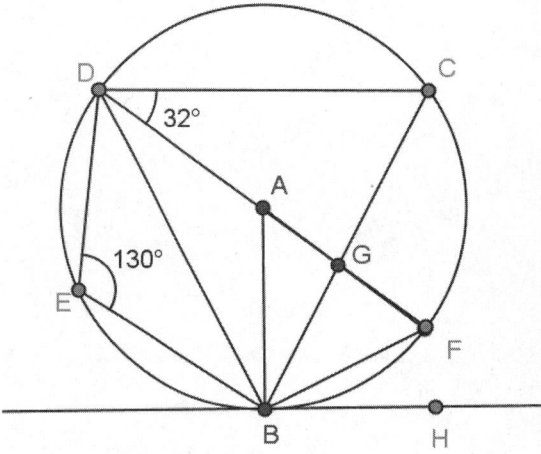

11. $\overleftrightarrow{BG}$ is tangent to circle A. $\overline{BE}$ is a diameter. Prove: (i) $f = a$ and (ii) $d = c$.

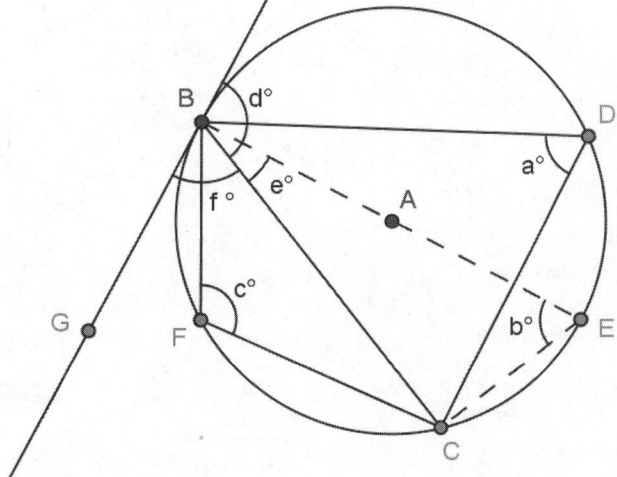

This page intentionally left blank

Lesson 14: Secant Lines; Secant Lines That Meet Inside a Circle

Classwork

Opening Exercise

$\overleftrightarrow{DB}$ is tangent to the circle as shown.

    a.   Find the values of a and b.

    b.   Is $\overline{CB}$ a diameter of the circle? Explain.

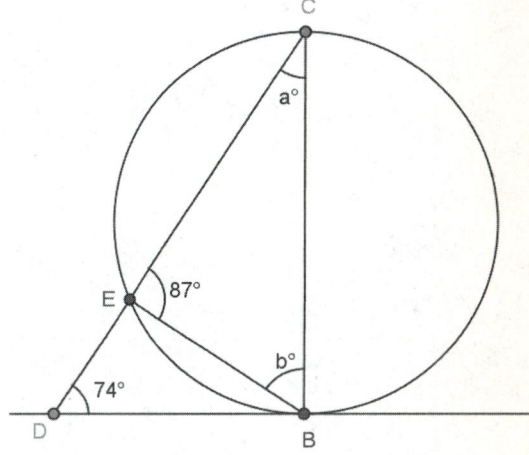

Exercises 1–2

1.   In circle P, $\overline{PO}$ is a radius, and $m\widehat{MO} = 142°$. Find $m\angle MOP$, and explain how you know.

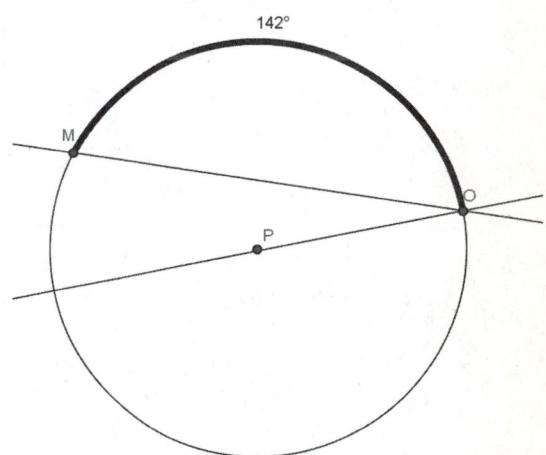

2. In the circle shown, $m\widehat{CE} = 55°$. Find $m\angle DEF$ and $m\widehat{EG}$. Explain your answer.

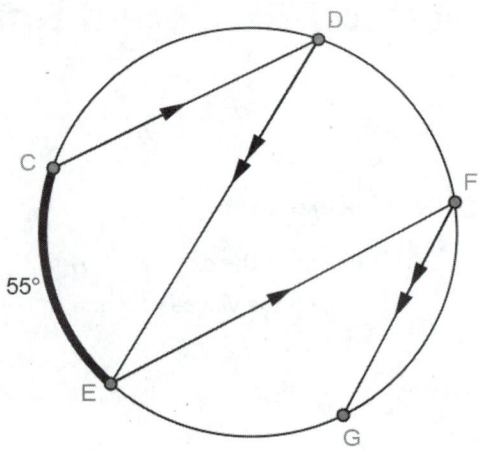

Example

a. Find x. Justify your answer.

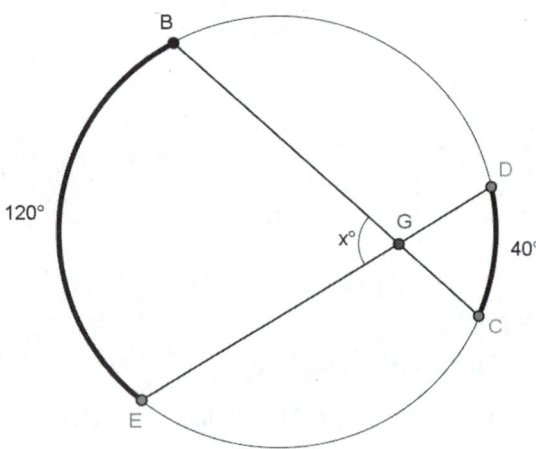

b. Find x.

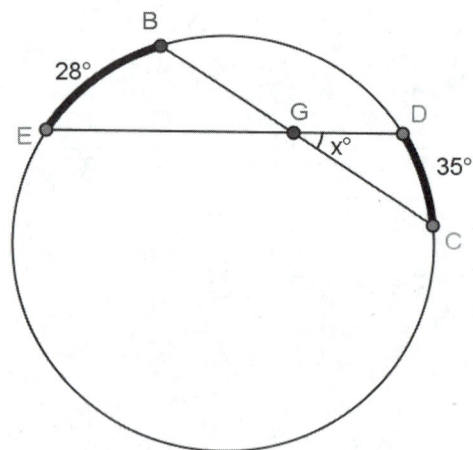

We can state the results of part (b) of this example as the following theorem:

SECANT ANGLE THEOREM—INTERIOR CASE: The measure of an angle whose vertex lies in the interior of a circle is equal to half the sum of the angle measures of the arcs intercepted by it and its vertical angle.

Exercises 3–7

In Exercises 3–5, find x and y.

3.

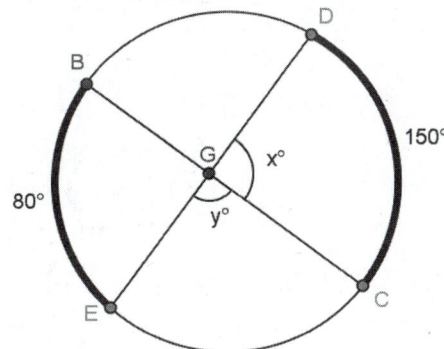

4.

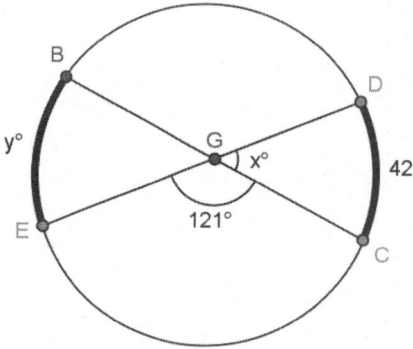

5.

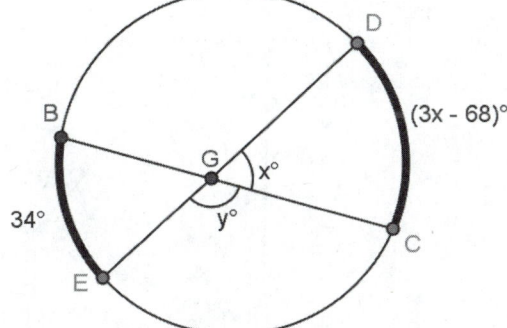

© 2015 Great Minds. eureka-math.org
GEO-M3-SE-B2-1.3.0-10.2015

6. In the circle shown, $\overline{BC}$ is a diameter. Find x and y.

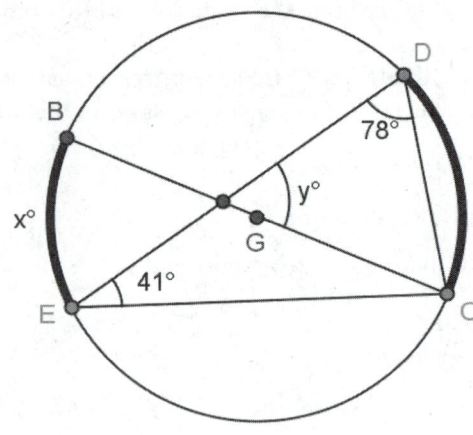

7. In the circle shown, $\overline{BC}$ is a diameter. $DC:BE = 2:1$. Prove $y = 180 - \dfrac{3}{2}x$ using a two-column proof.

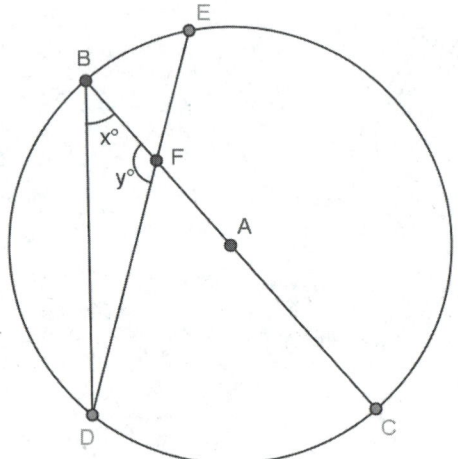

EUREKA
MATH™

Lesson Summary

THEOREM:

- **SECANT ANGLE THEOREM—INTERIOR CASE:** The measure of an angle whose vertex lies in the interior of a circle is equal to half the sum of the angle measures of the arcs intercepted by it and its vertical angle.

Relevant Vocabulary

- **TANGENT TO A CIRCLE:** A *tangent line to a circle* is a line in the same plane that intersects the circle in one and only one point. This point is called the *point of tangency*.
- **TANGENT SEGMENT/RAY:** A segment is a *tangent segment to a circle* if the line that contains it is tangent to the circle and one of the end points of the segment is a point of tangency. A ray is called a *tangent ray to a circle* if the line that contains it is tangent to the circle and the vertex of the ray is the point of tangency.
- **SECANT TO A CIRCLE:** A *secant line to a circle* is a line that intersects a circle in exactly two points.

Problem Set

In Problems 1–4, find x.

1.

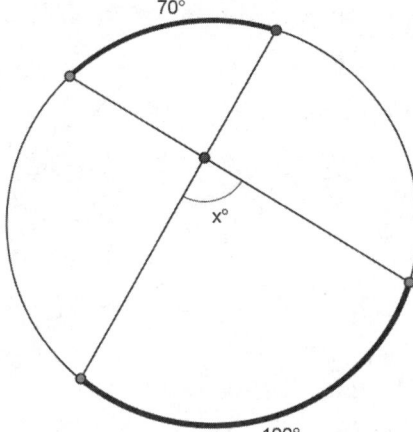

2.

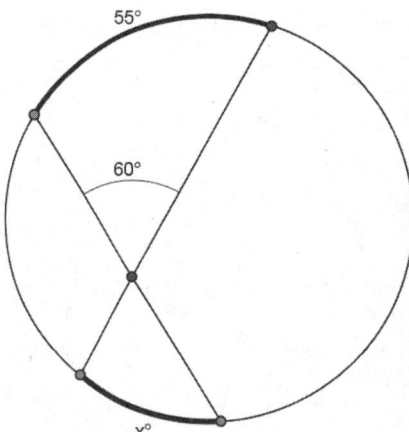

3.

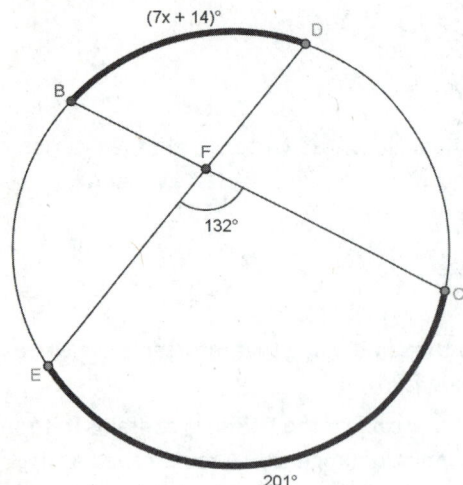

4.

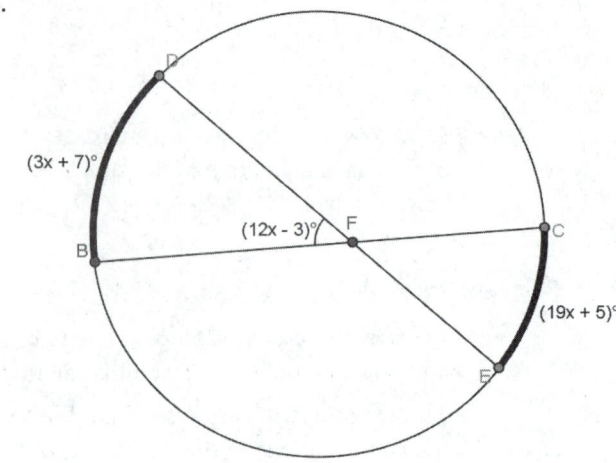

5.   Find x ($m\widehat{CE}$) and y ($m\widehat{DG}$).

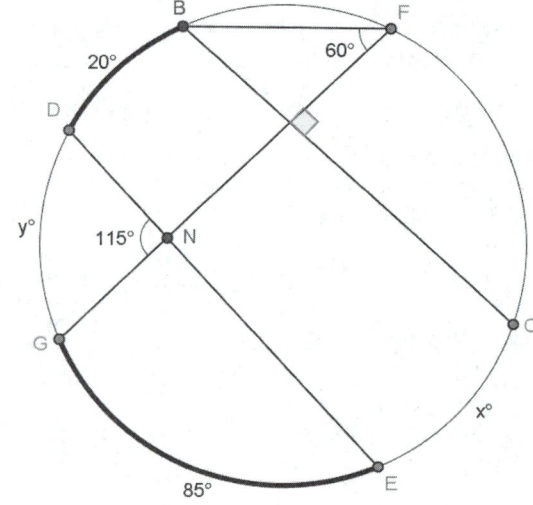

6.   Find the ratio of $m\widehat{EC} : m\widehat{DB}$.

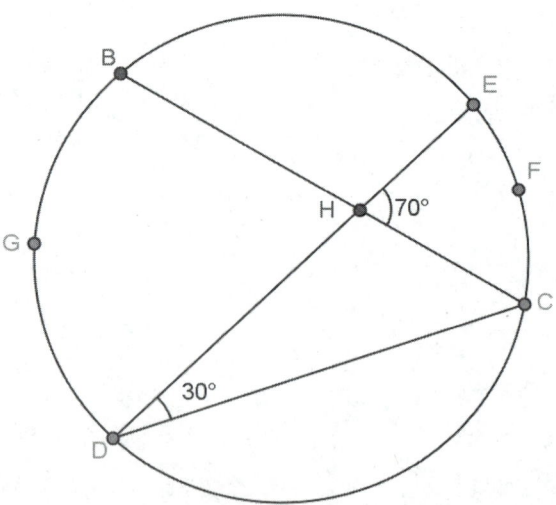

EUREKA
MATH

7. $\overline{BC}$ is a diameter of circle A. Find x.

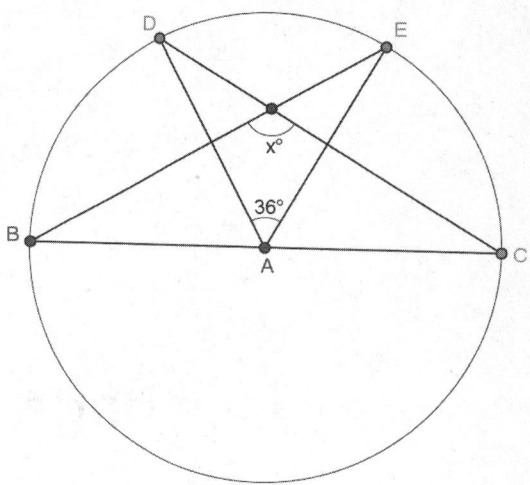

8. Show that the general formula we discovered in Example 1 also works for central angles. (Hint: Extend the radii to form two diameters, and use relationships between central angles and arc measure.)

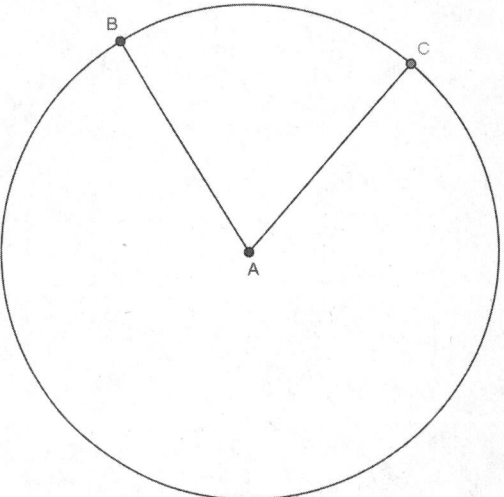

This page intentionally left blank

Lesson 15: Secant Angle Theorem, Exterior Case

Classwork

Opening Exercise

1. Shown below are circles with two intersecting secant chords.

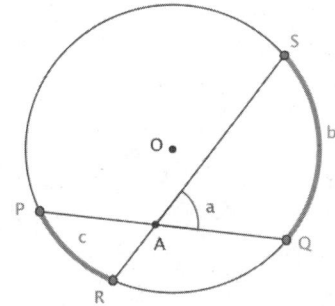 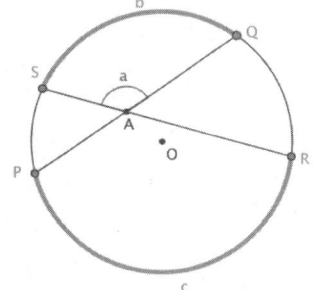

Measure a, b, and c in the two diagrams. Make a conjecture about the relationship between them.

a	b	c

CONJECTURE about the relationship between a, b, and c:

EUREKA MATH™

2. We will prove the following.

    SECANT ANGLE THEOREM—INTERIOR CASE: The measure of an angle whose vertex lies in the interior of a circle is equal to half the sum of the angle measures of the arcs intercepted by it and its vertical angle.

    We can interpret this statement in terms of the diagram below. Let b and c be the angle measures of the arcs intercepted by $\angle SAQ$ and $\angle PAR$. Then measure a is the average of b and c; that is, $a = \dfrac{b+c}{2}$.

    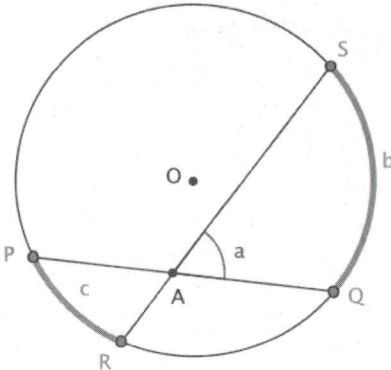

    a. Find as many pairs of congruent angles as you can in the diagram below. Express the measures of the angles in terms of b and c whenever possible.

    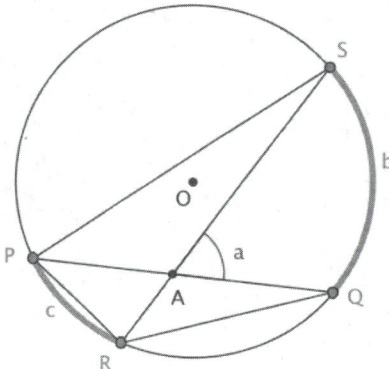

    b. Which triangles in the diagram are similar? Explain how you know.

    c. See if you can use one of the triangles to prove the secant angle theorem, interior case. (Hint: Use the exterior angle theorem.)

Lesson 15:   Secant Angle Theorem, Exterior Case

EUREKA
MATH™

Exploratory Challenge

Shown below are two circles with two secant chords intersecting outside the circle.

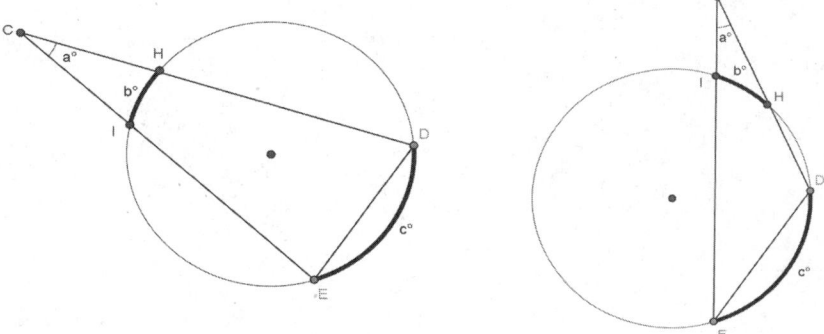

Measure a, b, and c. Make a conjecture about the relationship between them.

a	b	c

Conjecture about the relationship between a, b, and c:

Test your conjecture with another diagram.

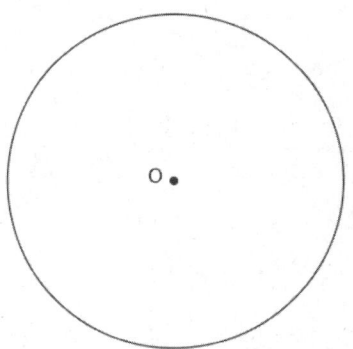

Exercises

Find x, y, and/or z.

1.

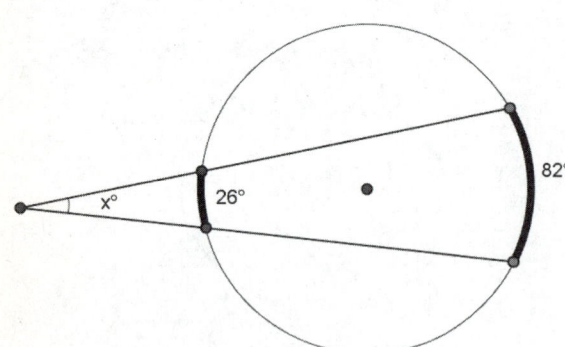

2.

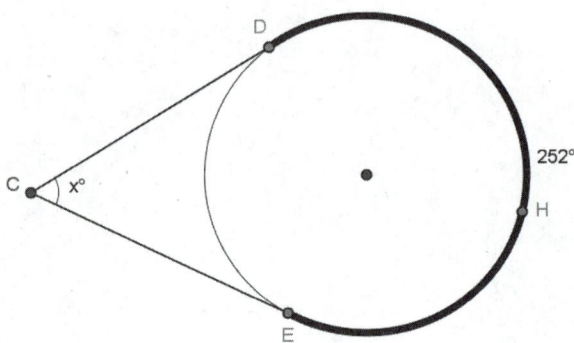

3.

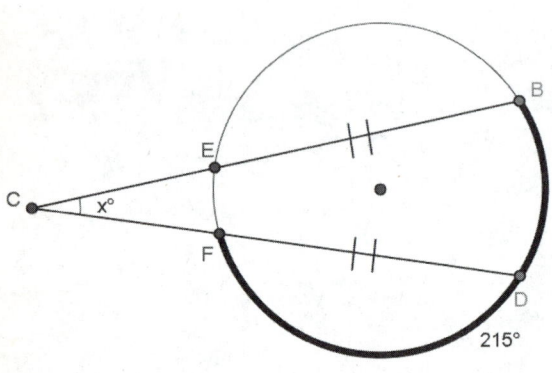

4.

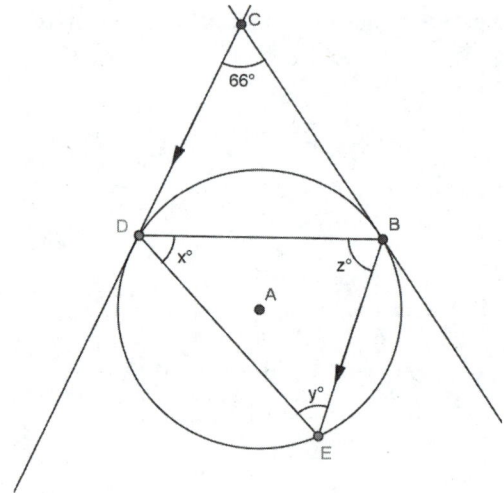

Closing Exercise

We have just developed proofs for an entire family of theorems. Each theorem in this family deals with two shapes and how they overlap. The two shapes are two intersecting lines and a circle.

In this exercise, you will summarize the different cases.

The Inscribed Angle Theorem and its Family of Theorems

Diagram	How the Two Shapes Overlap	Relationship between a, b, c, and d
(Inscribed Angle Theorem)		
(Secant–Tangent)		

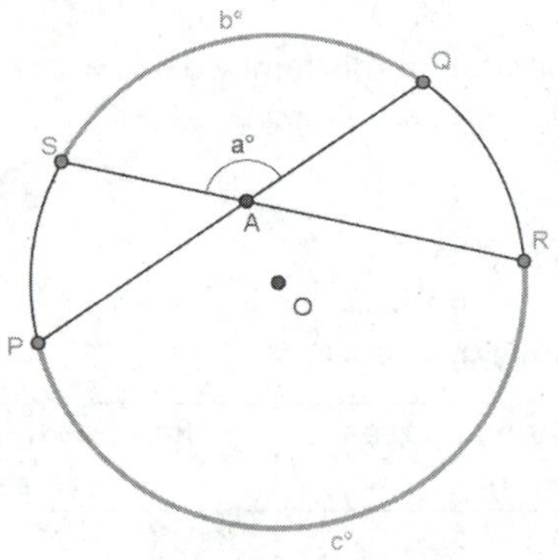

(Secant Angle Theorem—Interior)

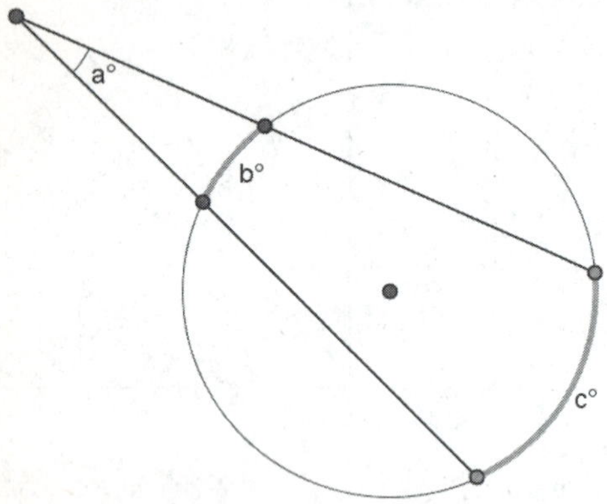

(Secant Angle Theorem—Exterior)

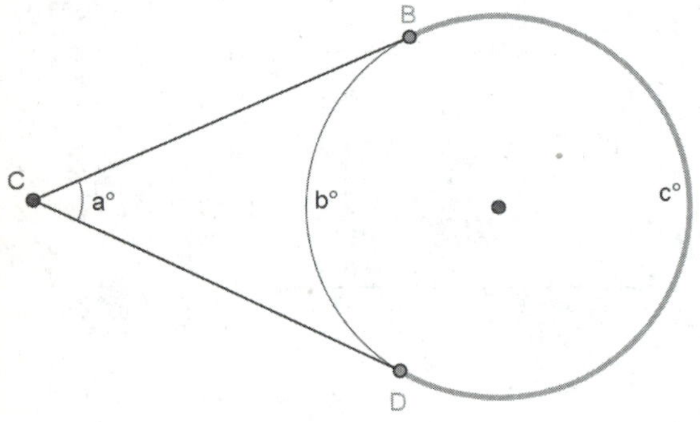

(Two Tangent Lines)

Lesson Summary

THEOREMS:

- **SECANT ANGLE THEOREM—INTERIOR CASE:** The measure of an angle whose vertex lies in the interior of a circle is equal to half the sum of the angle measures of the arcs intercepted by it and its vertical angle.

- **SECANT ANGLE THEOREM—EXTERIOR CASE:** The measure of an angle whose vertex lies in the exterior of the circle, and each of whose sides intersect the circle in two points, is equal to half the difference of the angle measures of its larger and smaller intercepted arcs.

Relevant Vocabulary

SECANT TO A CIRCLE: A *secant line to a circle* is a line that intersects a circle in exactly two points.

Problem Set

1. Find x.

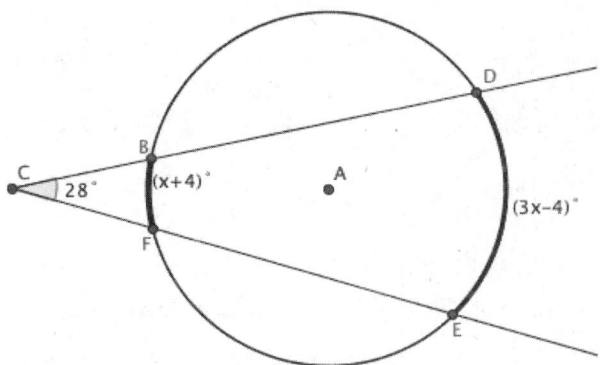

2. Find $m\angle DFE$ and $m\angle DGB$.

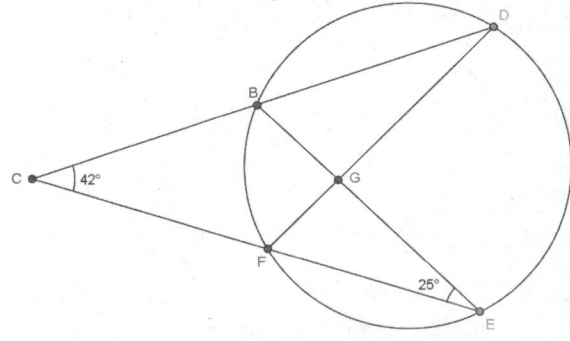

---

3. Find $m\angle ECD$, $m\angle DBE$, and $m\angle DEB$.

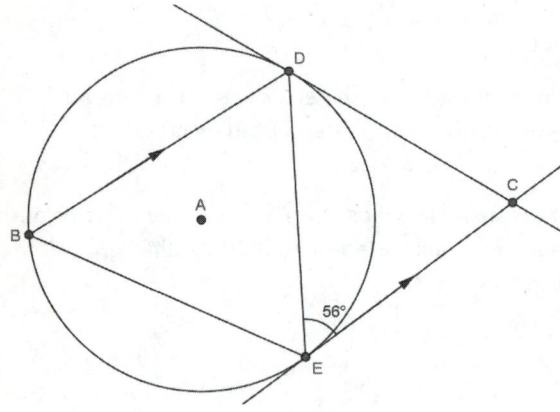

4. Find $m\angle FGE$ and $m\angle FHE$.

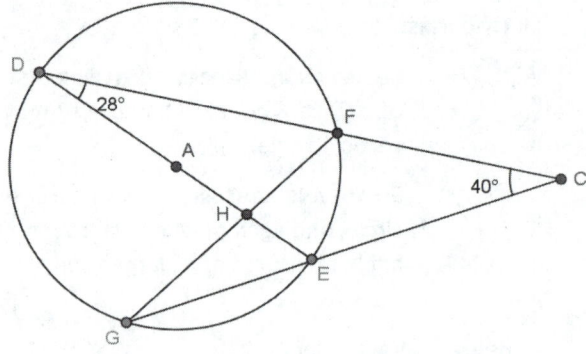

5. Find x and y.

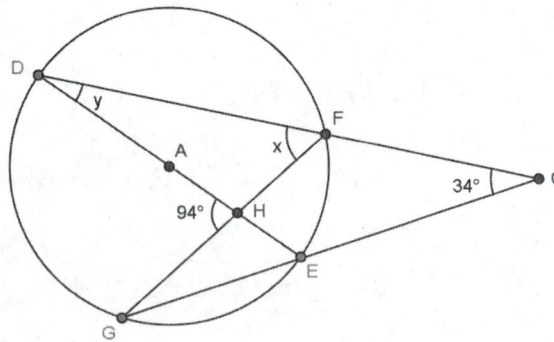

6. The radius of circle A is 4. $\overline{DC}$ and $\overline{CE}$ are tangent to the circle with $DC = 12$. Find $m\widehat{DE}$ and the area of quadrilateral $DAEC$ rounded to the nearest hundredth.

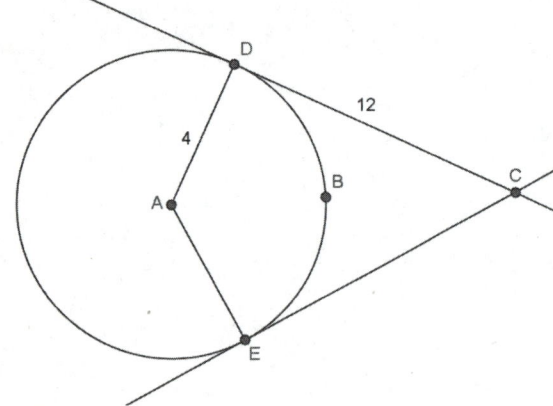

EUREKA
MATH™

7. Find the measure of $\overset{\frown}{BG}$, $\overset{\frown}{FB}$, and $\overset{\frown}{GF}$.

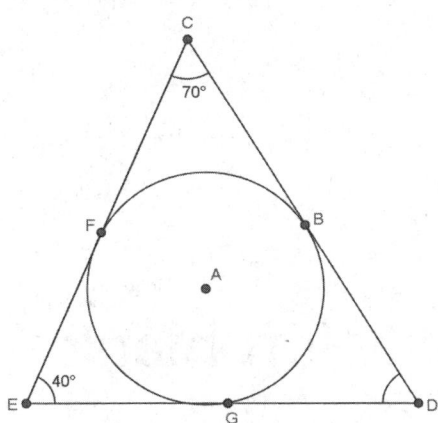

8. Find the values of x and y.

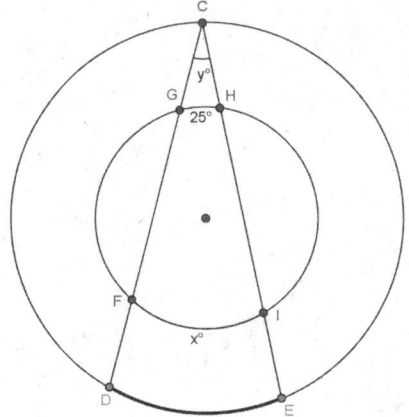

9. The radius of a circle is 6.

   a. If the angle formed between two tangent lines to the circle is 60°, how long are the segments between the point of intersection of the tangent lines and the circle?

   b. If the angle formed between the two tangent lines is 120°, how long are each of the segments between the point of intersection of the tangent lines and the point of tangency? Round to the nearest hundredth.

10. $\overline{DC}$ and $\overline{EC}$ are tangent to circle A. Prove $BD = BE$.

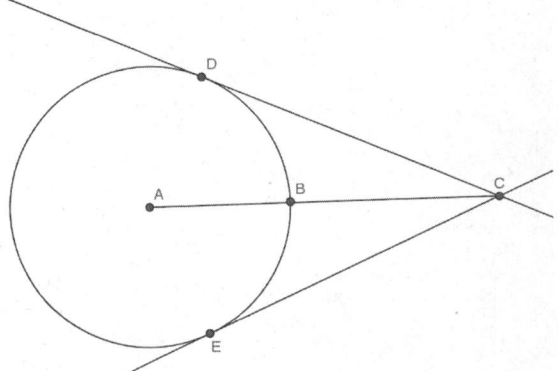

This page intentionally left blank

Lesson 16: Similar Triangles in Circle-Secant (or Circle-Secant-Tangent) Diagrams

Classwork

Opening Exercise

Identify the type of angle and the angle/arc relationship, and then find the measure of x.

a.

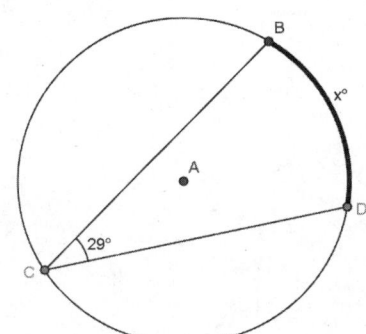

b.

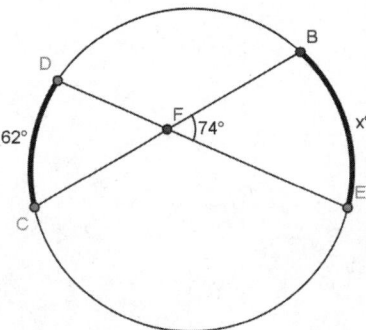

c.

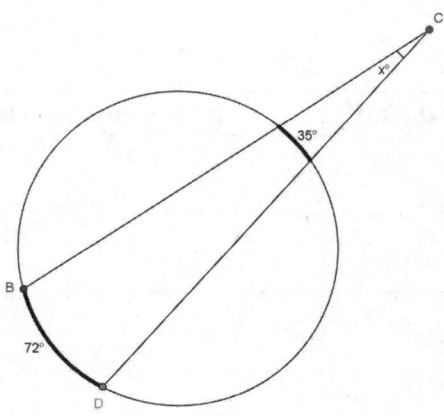

d.

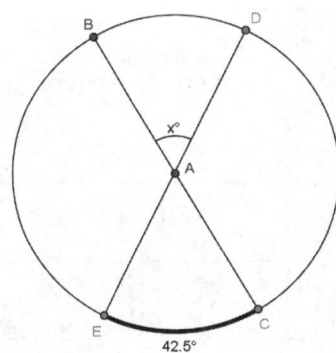

EUREKA
MATH™

Exploratory Challenge 1

Measure the lengths of the chords in centimeters, and record them in the table.

A.

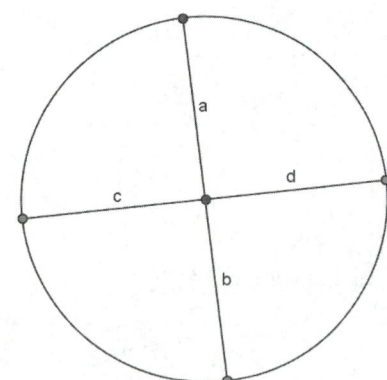

B.

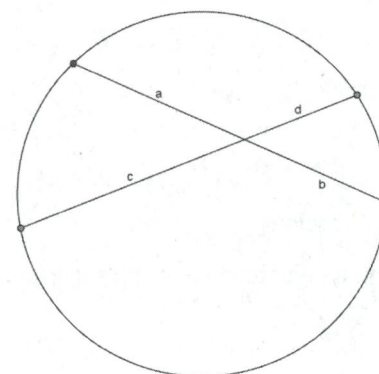

C.

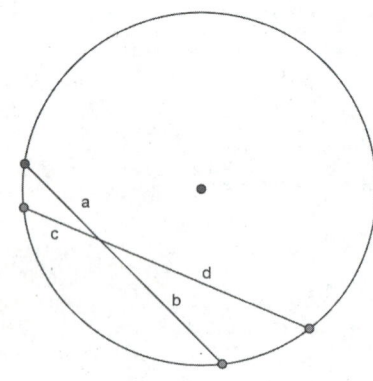

D.

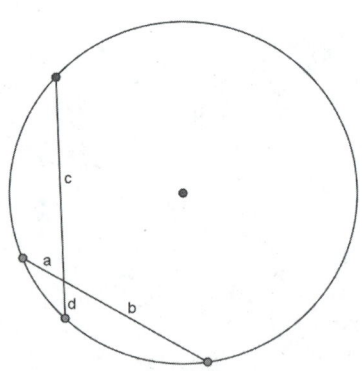

Circle	a (cm)	b (cm)	c (cm)	d (cm)	Do you notice a relationship?
A					
B					
C					
D					

Lesson 16:   Similar Triangles in Circle-Secant (or Circle-Secant-Tangent) Diagrams

EUREKA
MATH™

Exploratory Challenge 2

Measure the lengths of the chords in centimeters, and record them in the table.

A.

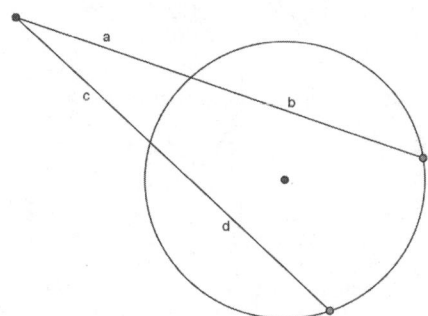

B.

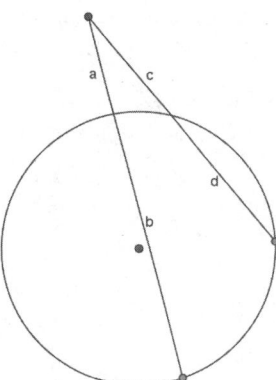

C.

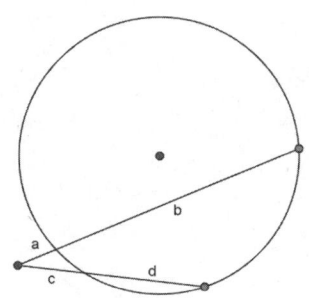

D.

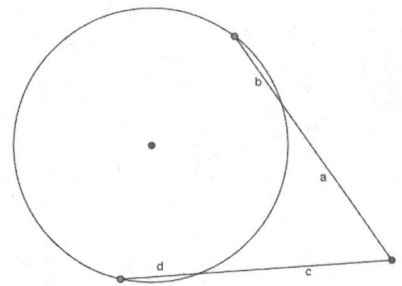

Circle	a (cm)	b (cm)	c (cm)	d (cm)	Do you notice a relationship?
A					
B					
C					
D					

© 2015 Great Minds. eureka-math.org
GEO-M3-SE-B2-1.3.0-10.2015

The Inscribed Angle Theorem and Its Family

Diagram	How the two shapes overlap	Relationship between a, b, c, and d

Lesson 16:   Similar Triangles in Circle-Secant (or Circle-Secant-Tangent) Diagrams

Lesson Summary

THEOREMS:

- When secant lines intersect inside a circle, use $a \cdot b = c \cdot d$.

- When secant lines intersect outside of a circle, use $a(a + b) = c(c + d)$.

- When a tangent line and a secant line intersect outside of a circle, use $a^2 = b(b + c)$.

Relevant Vocabulary

SECANT TO A CIRCLE: A *secant line to a circle* is a line that intersects a circle in exactly two points.

Problem Set

1. Find x.

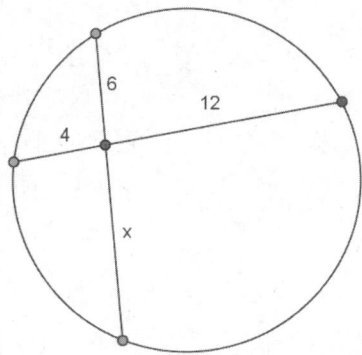

2. Find x.

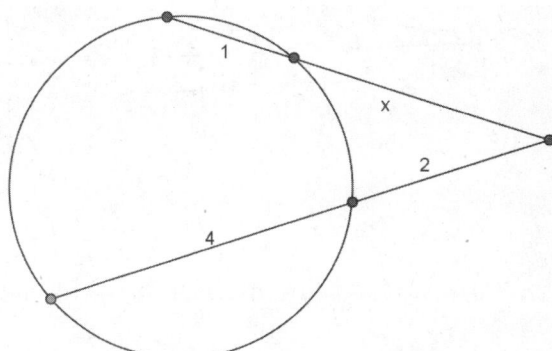

3. $DF < FB$, $DF \neq 1$, $DF < FE$, and all values are integers; prove $DF = 3$.

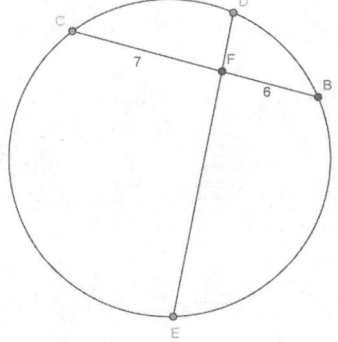

4. $CE = 6$, $CB = 9$, and $CD = 18$. Show $CF = 3$.

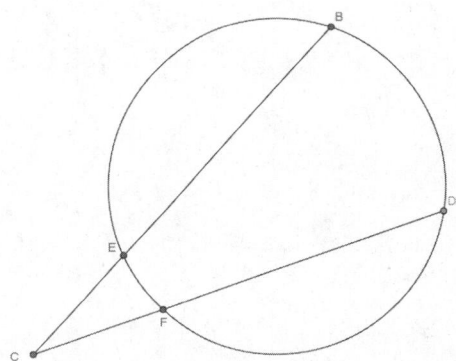

5. Find x.

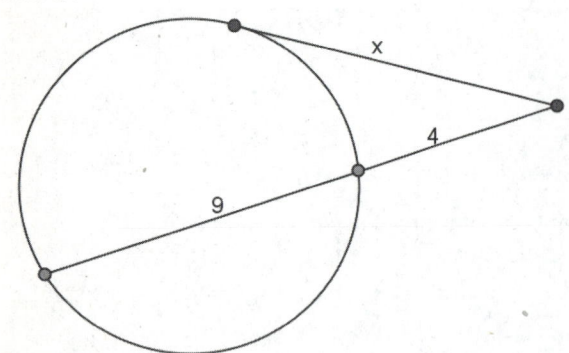

6. Find x.

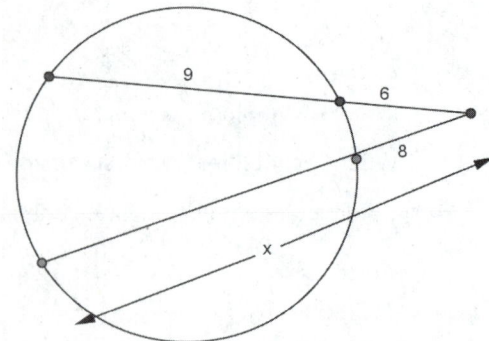

7. Find x.

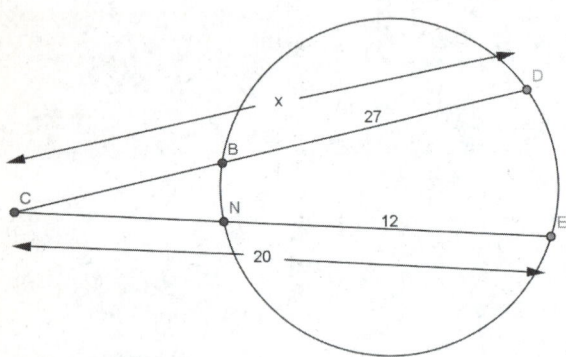

8. Find x.

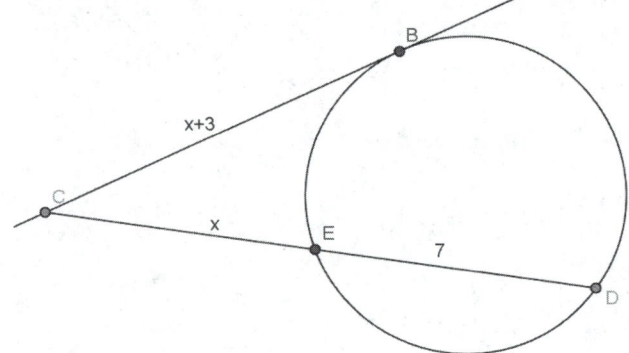

9. In the circle shown, $DE = 11$, $BC = 10$, and $DF = 8$. Find FE, BF, FC.

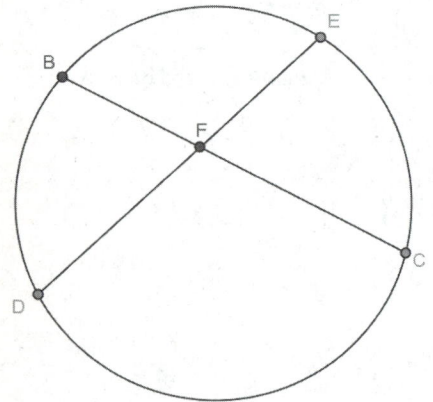

© 2015 Great Minds. eureka-math.org
GEO-M3-SE-B2-1.3.0-10.2015

EUREKA
MATH™

10. In the circle shown, $m\widehat{DBG} = 150°$, $m\widehat{DB} = 30°$, $m\angle CEF = 60°$, $DF = 8$, $DB = 4$, and $GF = 12$.

   a. Find $m\angle GDB$.

   b. Prove $\triangle DBF \sim \triangle ECF$.

   c. Set up a proportion using $\overline{CE}$ and $\overline{GE}$.

   d. Set up an equation with CE and GE using a theorem for segment lengths from this section.

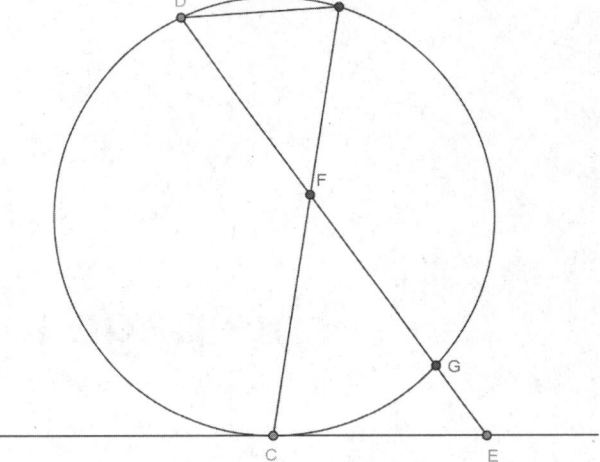

This page intentionally left blank

Lesson 17: Writing the Equation for a Circle

Exercises 1–2

1. What is the length of the segment shown on the coordinate plane below?

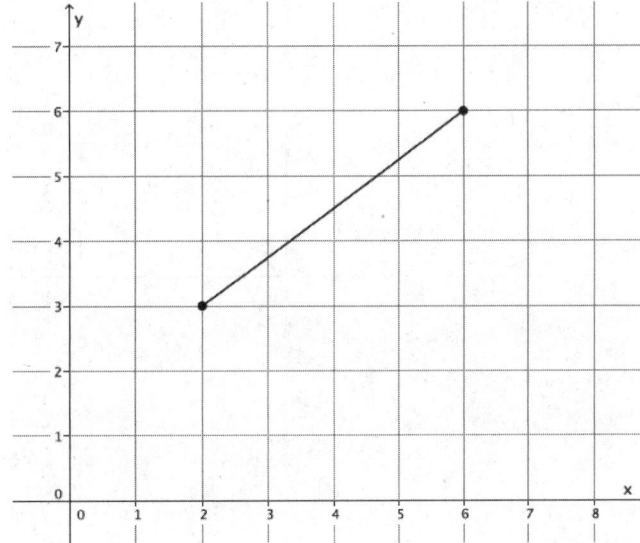

2. Use the distance formula to determine the distance between points $(9, 15)$ and $(3, 7)$.

Example 1

If we graph all of the points whose distance from the origin is equal to 5, what shape will be formed?

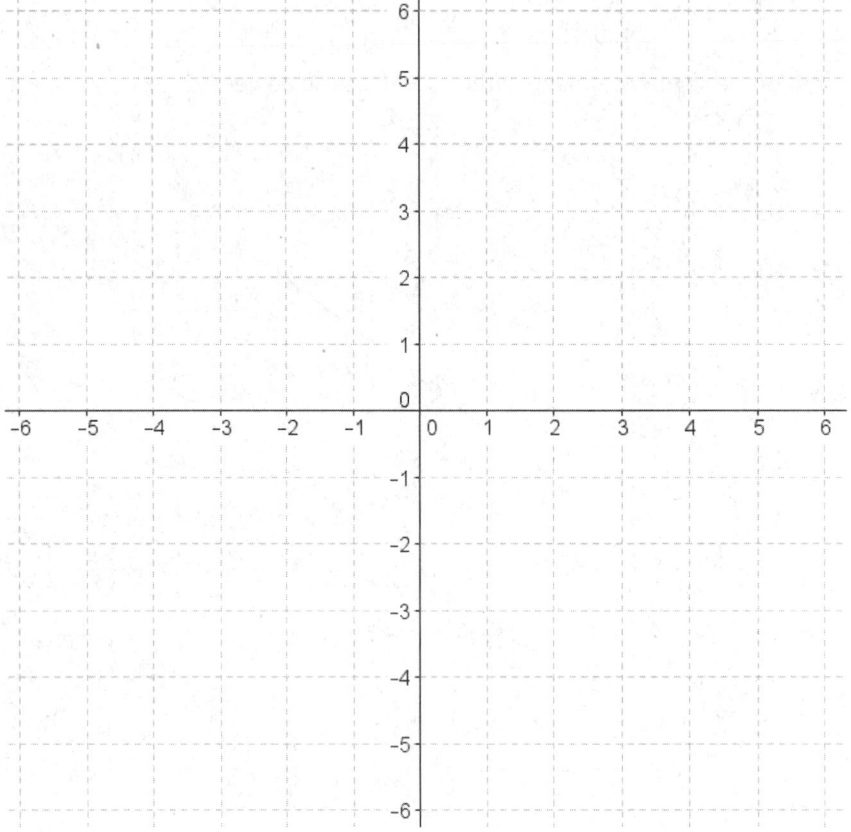

EUREKA
MATH™

Example 2

Let's look at another circle, one whose center is not at the origin. Shown below is a circle with center $(2, 3)$ and radius 5.

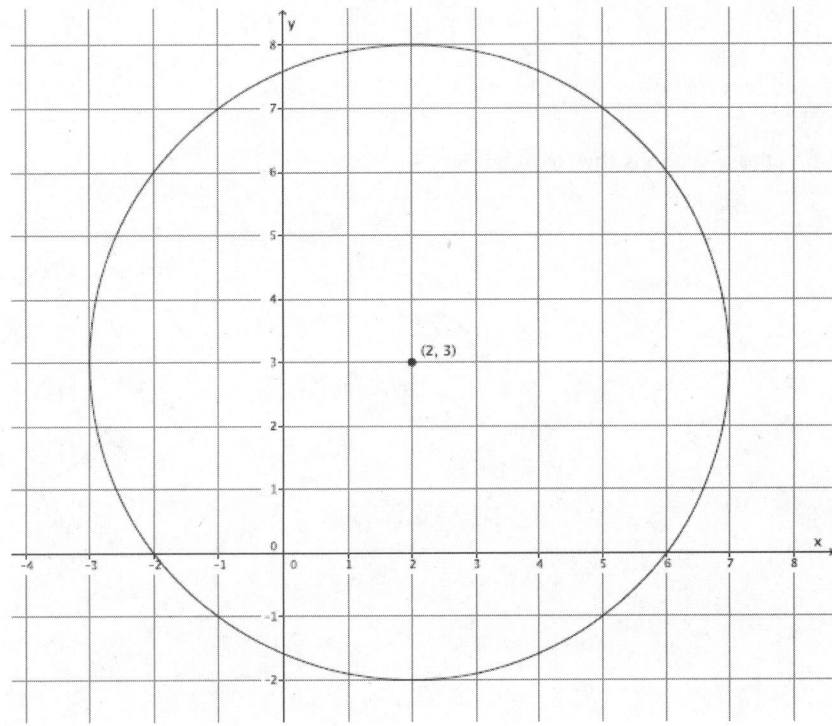

Exercises 3–11

3.   Write an equation for the circle whose center is at $(9, 0)$ and has radius 7.

4.   Write an equation whose graph is the circle below.

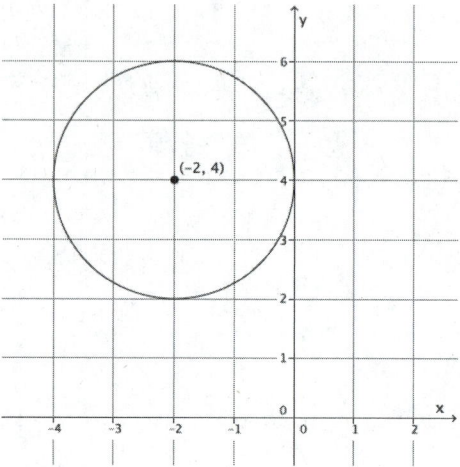

5.   What is the radius and center of the circle given by the equation $(x + 12)^2 + (y - 4)^2 = 81$?

6.   Petra is given the equation $(x - 15)^2 + (y + 4)^2 = 100$ and identifies its graph as a circle whose center is $(-15, 4)$ and radius is 10. Has Petra made a mistake? Explain.

Lesson 17:   Writing the Equation for a Circle

EUREKA MATH™

7.

    a.   What is the radius of the circle with center $(3, 10)$ that passes through $(12, 12)$?

    b.   What is the equation of this circle?

8.   A circle with center $(2, -5)$ is tangent to the x-axis.

    a.   What is the radius of the circle?

    b.   What is the equation of the circle?

9.   Two points in the plane, $A(-3, 8)$ and $B(17, 8)$, represent the endpoints of the diameter of a circle.

    a.   What is the center of the circle? Explain.

    b.   What is the radius of the circle? Explain.

c.   Write the equation of the circle.

10. Consider the circles with the following equations:

$$x^2 + y^2 = 25 \text{ and}$$

$$(x - 9)^2 + (y - 12)^2 = 100.$$

a.   What are the radii of the circles?

b.   What is the distance between the centers of the circles?

c.   Make a rough sketch of the two circles to explain why the circles must be tangent to one another.

© 2015 Great Minds. eureka-math.org
GEO-M3-SE-B2-1.3.0-10.2015

11. A circle is given by the equation $(x^2 + 2x + 1) + (y^2 + 4y + 4) = 121$.

    a.   What is the center of the circle?

    b.   What is the radius of the circle?

    c.   Describe what you had to do in order to determine the center and the radius of the circle.

© 2015 Great Minds. eureka-math.org
GEO-M3-SE-B2-1.3.0-10.2015

Lesson Summary

$(x - a)^2 + (y - b)^2 = r^2$ is the center-radius form of the general equation for any circle with radius r and center (a, b).

Problem Set

1. Write the equation for a circle with center $\left(\frac{1}{2}, \frac{3}{7}\right)$ and radius $\sqrt{13}$.

2. What is the center and radius of the circle given by the equation $x^2 + (y - 11)^2 = 144$?

3. A circle is given by the equation $x^2 + y^2 = 100$. Which of the following points are on the circle?
    a. $(0, 10)$
    b. $(-8, 6)$
    c. $(-10, -10)$
    d. $(45, 55)$
    e. $(-10, 0)$

4. Determine the center and radius of each circle.
    a. $3x^2 + 3y^2 = 75$
    b. $2(x + 1)^2 + 2(y + 2)^2 = 10$
    c. $4(x - 2)^2 + 4(y - 9)^2 - 64 = 0$

5. A circle has center $(-13, \pi)$ and passes through the point $(2, \pi)$.
    a. What is the radius of the circle?
    b. Write the equation of the circle.

6. Two points in the plane, $A(19, 4)$ and $B(19, -6)$, represent the endpoints of the diameter of a circle.
    a. What is the center of the circle?
    b. What is the radius of the circle?
    c. Write the equation of the circle.

EUREKA
MATH

7. Write the equation of the circle shown below.

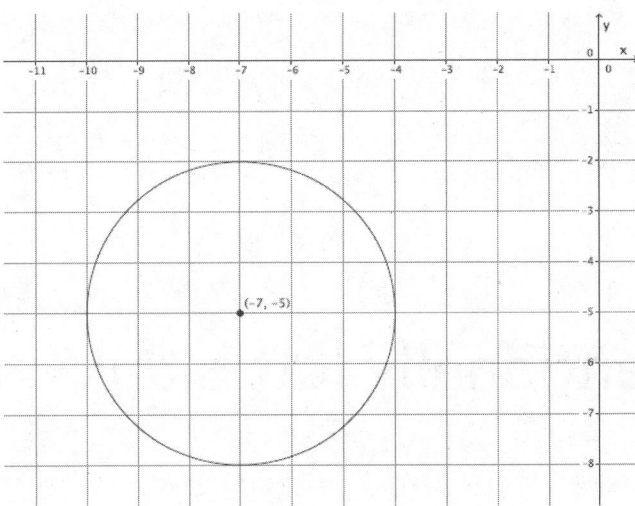

8. Write the equation of the circle shown below.

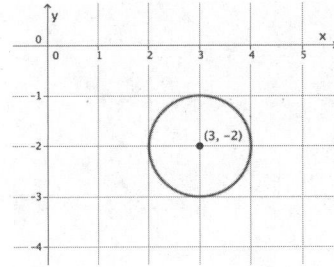

9. Consider the circles with the following equations:

$$x^2 + y^2 = 2 \text{ and}$$

$$(x - 3)^2 + (y - 3)^2 = 32.$$

   a. What are the radii of the two circles?

   b. What is the distance between their centers?

   c. Make a rough sketch of the two circles to explain why the circles must be tangent to one another.

This page intentionally left blank

Lesson 18: Recognizing Equations of Circles

Opening Exercise

a.   Express this as a trinomial: $(x - 5)^2$.

b.   Express this as a trinomial: $(x + 4)^2$.

c.   Factor the trinomial: $x^2 + 12x + 36$.

d.   Complete the square to solve the following equation: $x^2 + 6x = 40$.

EUREKA
MATH™

Lesson 18:   Recognizing Equations of Circles

S.143

© 2015 Great Minds. eureka-math.org
GEO-M3-SE-B2-1.3.0-10.2015

Example 1

The following is the equation of a circle with radius 5 and center $(1, 2)$. Do you see why?

$$x^2 - 2x + 1 + y^2 - 4y + 4 = 25$$

Exercise 1

1. Rewrite the following equations in the form $(x - a)^2 + (y - b)^2 = r^2$.

   a. $x^2 + 4x + 4 + y^2 - 6x + 9 = 36$

   b. $x^2 - 10x + 25 + y^2 + 14y + 49 = 4$

Example 2

What is the center and radius of the following circle?

$$x^2 + 4x + y^2 - 12y = 41$$

Exercises 2–4

2. Identify the center and radius for each of the following circles.

a. $x^2 - 20x + y^2 + 6y = 35$

b. $x^2 - 3x + y^2 - 5y = \dfrac{19}{2}$

3. Could the circle with equation $x^2 - 6x + y^2 - 7 = 0$ have a radius of 4? Why or why not?

4. Stella says the equation $x^2 - 8x + y^2 + 2y = 5$ has a center of $(4, -1)$ and a radius of 5. Is she correct? Why or why not?

Example 3

Could $x^2 + y^2 + Ax + By + C = 0$ represent a circle?

EUREKA
MATH

© 2015 Great Minds. eureka-math.org
GEO-M3-SE-B2-1.3.0-10.2015

Exercise 5

5. Identify the graphs of the following equations as a circle, a point, or an empty set.

    a. $x^2 + y^2 + 4x = 0$

    b. $x^2 + y^2 + 6x - 4y + 15 = 0$

    c. $2x^2 + 2y^2 - 5x + y + \frac{13}{4} = 0$

Problem Set

1. Identify the centers and radii of the following circles.
    a. $(x + 25)^2 + y^2 = 1$
    b. $x^2 + 2x + y^2 - 8y = 8$
    c. $x^2 - 20x + y^2 - 10y + 25 = 0$
    d. $x^2 + y^2 = 19$
    e. $x^2 + x + y^2 + y = -\frac{1}{4}$

2. Sketch graphs of the following equations.
    a. $x^2 + y^2 + 10x - 4y + 33 = 0$
    b. $x^2 + y^2 + 14x - 16y + 104 = 0$
    c. $x^2 + y^2 + 4x - 10y + 29 = 0$

3. Chante claims that two circles given by $(x + 2)^2 + (y - 4)^2 = 49$ and $x^2 + y^2 - 6x + 16y + 37 = 0$ are externally tangent. She is right. Show that she is.

4. Draw a circle. Randomly select a point in the interior of the circle; label the point A. Construct the greatest radius circle with center A that lies within the circular region defined by the original circle. Hint: Draw a line through the center, the circle, and point A.

EUREKA
MATH™

Lesson 19: Equations for Tangent Lines to Circles

Classwork

Opening Exercise

A circle of radius 5 passes through points $A(-3, 3)$ and $B(3, 1)$.

    a.   What is the special name for segment AB?

    b.   How many circles can be drawn that meet the given criteria? Explain how you know.

    c.   What is the slope of $\overline{AB}$?

    d.   Find the midpoint of $\overline{AB}$.

    e.   Find the equation of the line containing a diameter of the given circle perpendicular to $\overline{AB}$.

    f.   Is there more than one answer possible for part (e)?

Example 1

Consider the circle with equation $(x - 3)^2 + (y - 5)^2 = 20$. Find the equations of two tangent lines to the circle that each has slope $-\frac{1}{2}$.

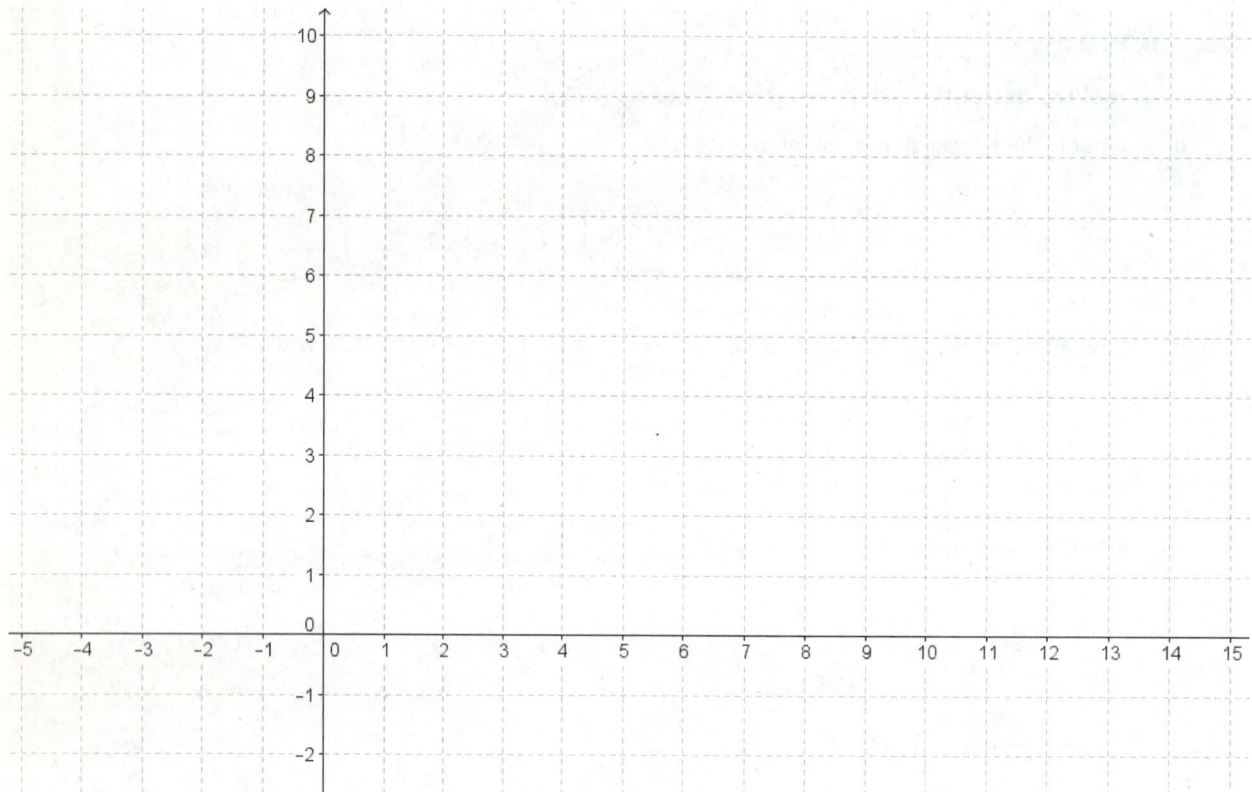

© 2015 Great Minds. eureka-math.org
GEO-M3-SE-B2-1.3.0-10.2015

Exercise 1

Consider the circle with equation $(x - 4)^2 + (y - 5)^2 = 20$. Find the equations of two tangent lines to the circle that each has slope 2.

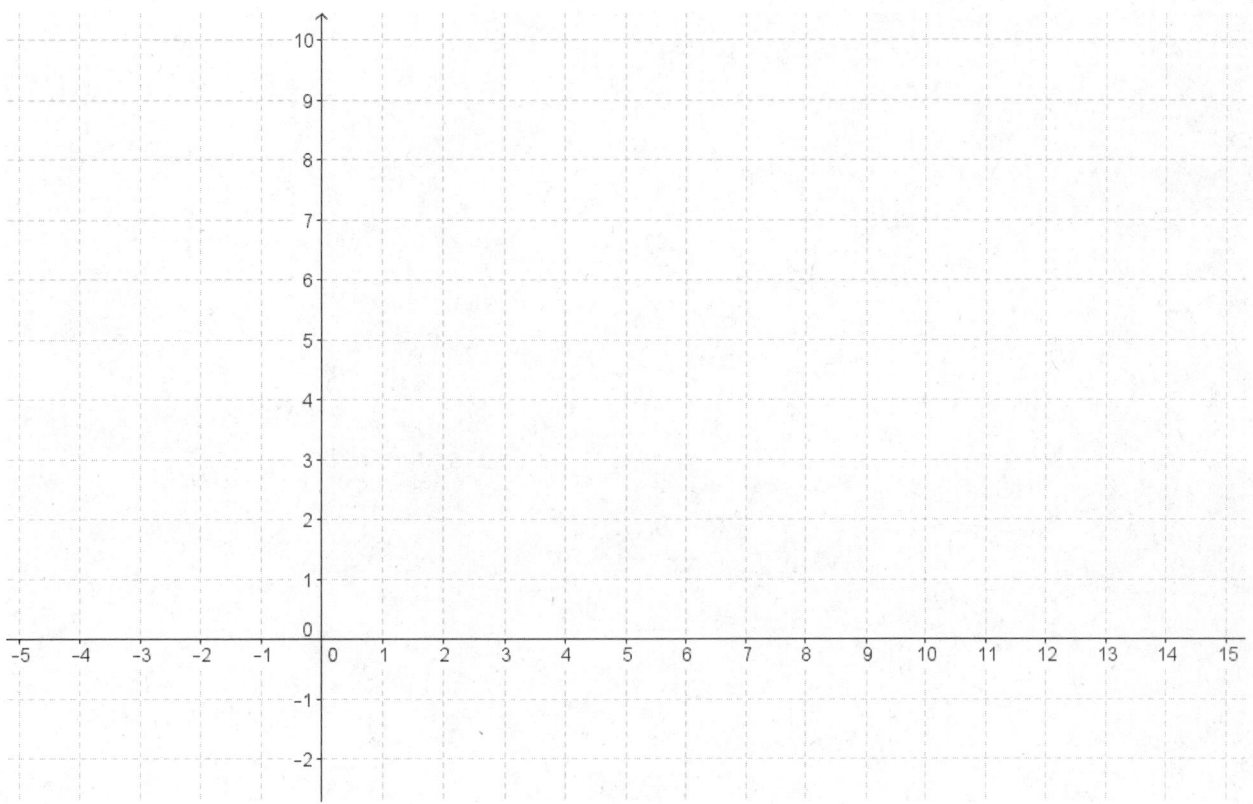

EUREKA
MATH™

Example 2

Refer to the diagram below.

Let $p > 1$. What is the equation of the tangent line to the circle $x^2 + y^2 = 1$ through the point $(p, 0)$ on the x-axis with a point of tangency in the upper half-plane?

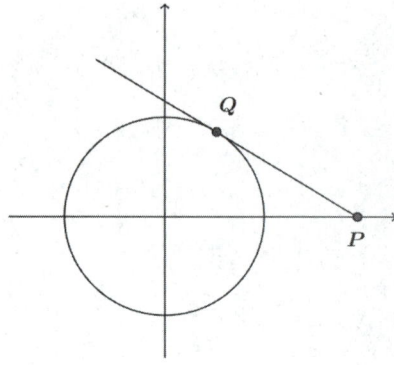

Exercises 2–4

2. Use the same diagram from Example 2 above, but label the point of tangency in the lower half-plane as Q'.

    a.   What are the coordinates of Q'?

    b.   What is the slope of $\overline{OQ'}$?

EUREKA
MATH™

    c.   What is the slope of $\overline{Q'P}$?

    d.   Find the equation of the second tangent line to the circle through $(p, 0)$.

3.   Show that a circle with equation $(x - 2)^2 + (y + 3)^2 = 160$ has two tangent lines with equations $y + 15 = \frac{1}{3}(x - 6)$ and $y - 9 = \frac{1}{3}(x + 2)$.

4.   Could a circle given by the equation $(x - 5)^2 + (y - 1)^2 = 25$ have tangent lines given by the equations $y - 4 = \frac{4}{3}(x - 1)$ and $y - 5 = \frac{3}{4}(x - 8)$? Explain how you know.

EUREKA
MATH™

Lesson 19:   Equations for Tangent Lines to Circles

S.153

© 2015 Great Minds. eureka-math.org
GEO-M3-SE-B2-1.3.0-10.2015

Problem Set

1. Consider the circle $(x - 1)^2 + (y - 2)^2 = 16$. There are two lines tangent to this circle having a slope of 0.
   a. Find the coordinates of the points of tangency.
   b. Find the equations of the two tangent lines.

2. Consider the circle $x^2 - 4x + y^2 + 10y + 13 = 0$. There are two lines tangent to this circle having a slope of $\frac{2}{3}$.
   a. Find the coordinates of the two points of tangency.
   b. Find the equations of the two tangent lines.

3. What are the coordinates of the points of tangency of the two tangent lines through the point $(1, 1)$ each tangent to the circle $x^2 + y^2 = 1$?

4. What are the coordinates of the points of tangency of the two tangent lines through the point $(-1, -1)$ each tangent to the circle $x^2 + y^2 = 1$?

5. What is the equation of the tangent line to the circle $x^2 + y^2 = 1$ through the point $(6, 0)$?

6. D'Andre said that a circle with equation $(x - 2)^2 + (y - 7)^2 = 13$ has a tangent line represented by the equation $y - 5 = -\frac{3}{2}(x + 1)$. Is he correct? Explain.

© 2015 Great Minds. eureka-math.org
GEO-M3-SE-B2-1.3.0-10.2015

7. Kamal gives the following proof that $y - 1 = \frac{8}{9}(x + 10)$ is the equation of a line that is tangent to a circle given by

$$(x + 1)^2 + (y - 9)^2 = 145.$$

The circle has center $(-1, 9)$ and radius 12.04. The point $(-10, 1)$ is on the circle because

$$(-10 + 1)^2 + (1 - 9)^2 = (-9)^2 + (-8)^2 = 145.$$

The slope of the radius is $\dfrac{9 - 1}{-1 + 10} = \dfrac{8}{9}$; therefore, the equation of the tangent line is $y - 1 = \frac{8}{9}(x + 10)$.

   a. Kerry said that Kamal has made an error. What was Kamal's error? Explain what he did wrong.

   b. What should the equation for the tangent line be?

8. Describe a similarity transformation that maps a circle given by $x^2 + 6x + y^2 - 2y = 71$ to a circle of radius 3 that is tangent to both axes in the first quadrant.

EUREKA MATH™

This page intentionally left blank

Lesson 20: Cyclic Quadrilaterals

Classwork

Opening Exercise

Given cyclic quadrilateral $ABCD$ shown in the diagram, prove that $x + y = 180°$.

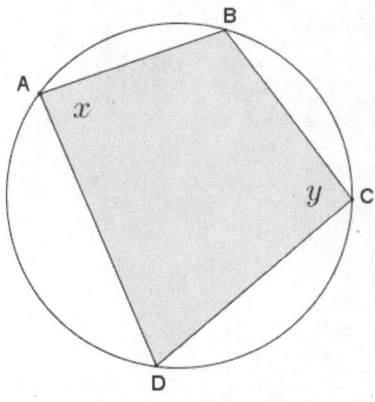

Example

Given quadrilateral $ABCD$ with $m\angle A + m\angle C = 180°$, prove that quadrilateral $ABCD$ is cyclic; in other words, prove that points A, B, C, and D lie on the same circle.

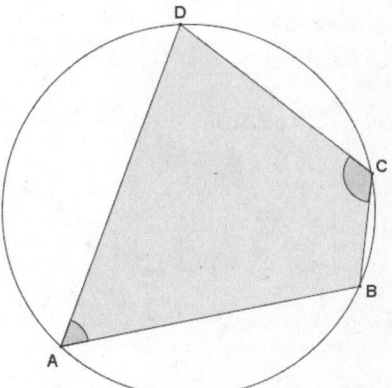

Exercises

1. Assume that vertex D'' lies inside the circle as shown in the diagram. Use a similar argument to Example 1 to show that vertex D'' cannot lie inside the circle.

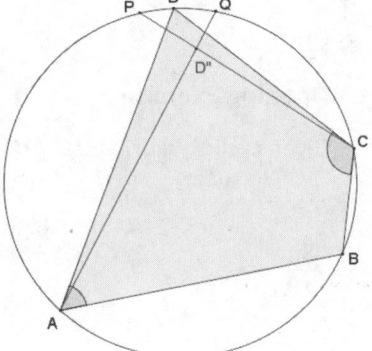

© 2015 Great Minds. eureka-math.org
GEO-M3-SE-B2-1.3.0-10.2015

EUREKA
MATH™

2. Quadrilateral $PQRS$ is a cyclic quadrilateral. Explain why $\triangle PQT \sim \triangle SRT$.

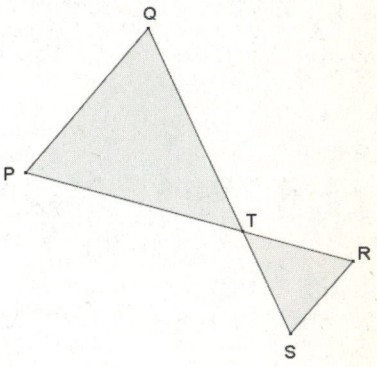

3. A cyclic quadrilateral has perpendicular diagonals. What is the area of the quadrilateral in terms of a, b, c, and d as shown?

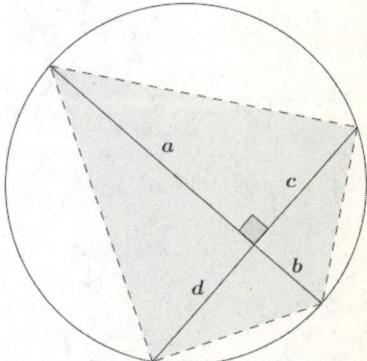

EUREKA
MATH™

Lesson 20: Cyclic Quadrilaterals

S.159

© 2015 Great Minds. eureka-math.org
GEO-M3-SE-B2-1.3.0-10.2015

4. Show that the triangle in the diagram has area $\frac{1}{2} ab \sin(w)$.

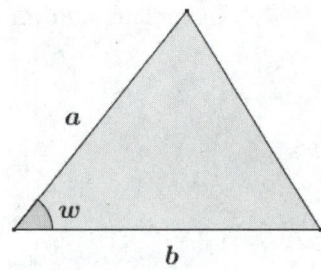

5. Show that the triangle with obtuse angle $(180 - w)°$ has area $\frac{1}{2} ab \sin(w)$.

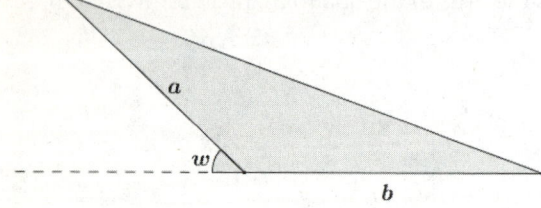

Lesson 20:   Cyclic Quadrilaterals

EUREKA
MATH™

6. Show that the area of the cyclic quadrilateral shown in the diagram is Area $= \frac{1}{2}(a + b)(c + d)\sin(w)$.

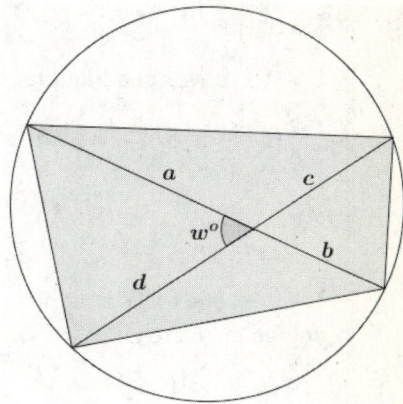

Lesson Summary

THEOREMS:

Given a convex quadrilateral, the quadrilateral is cyclic if and only if one pair of opposite angles is supplementary.

The area of a triangle with side lengths a and b and acute included angle with degree measure w:

$$\text{Area} = \frac{1}{2}ab \cdot \sin(w).$$

The area of a cyclic quadrilateral $ABCD$ whose diagonals $\overline{AC}$ and $\overline{BD}$ intersect to form an acute or right angle with degree measure w:

$$\text{Area}(ABCD) = \frac{1}{2} \cdot AC \cdot BD \cdot \sin(w).$$

Relevant Vocabulary

CYCLIC QUADRILATERAL: A quadrilateral inscribed in a circle is called a *cyclic quadrilateral*.

Problem Set

1. Quadrilateral $BDCE$ is cyclic, O is the center of the circle, and $m\angle BOC = 130°$. Find $m\angle BEC$.

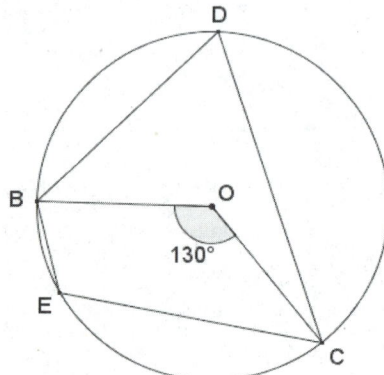

EUREKA
MATH™

2. Quadrilateral $FAED$ is cyclic, $AX = 8$, $FX = 6$, $XD = 3$, and $m\angle AXE = 130°$. Find the area of quadrilateral $FAED$.

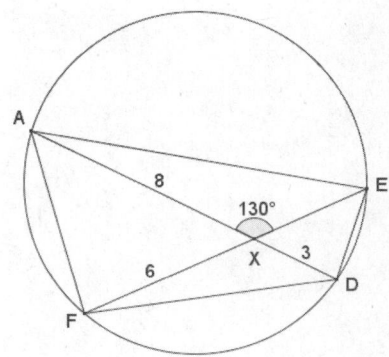

3. In the diagram below, $\overline{BE} \parallel \overline{CD}$ and $m\angle BED = 72°$. Find the value of s and t.

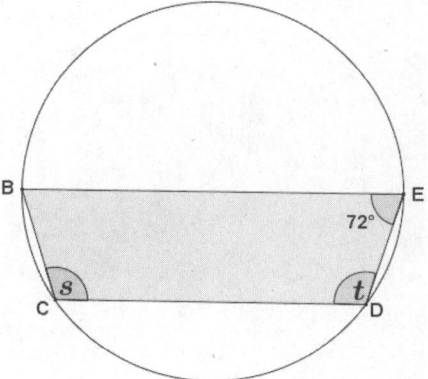

4. In the diagram below, $\overline{BC}$ is the diameter, $m\angle BCD = 25°$, and $\overline{CE} \cong \overline{DE}$. Find $m\angle CED$ and $m\angle EDC$.

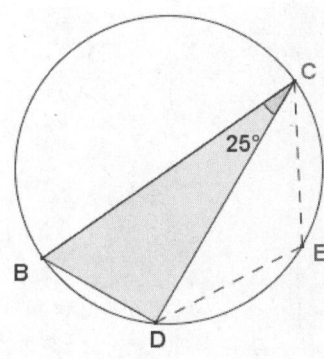

5. In circle A, $m\angle ABD = 15°$. Find $m\angle BCD$.

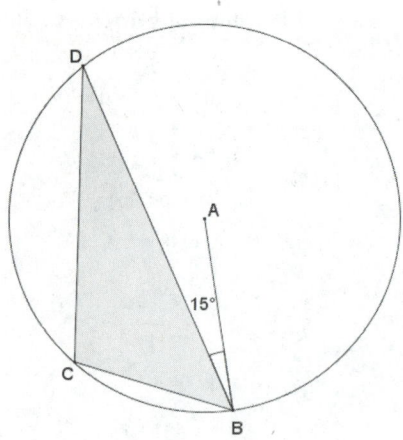

6. Given the diagram below, O is the center of the circle. If $m\angle NOP = 112°$, find $m\angle PQE$.

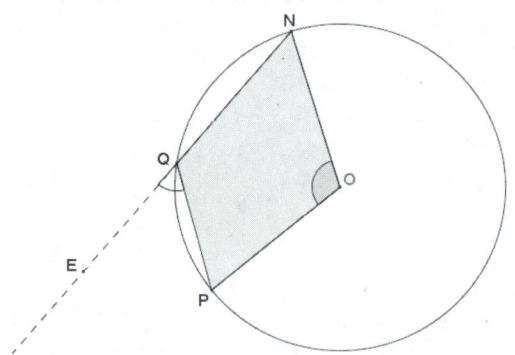

7. Given the angle measures as indicated in the diagram below, prove that vertices C, B, E, and D lie on a circle.

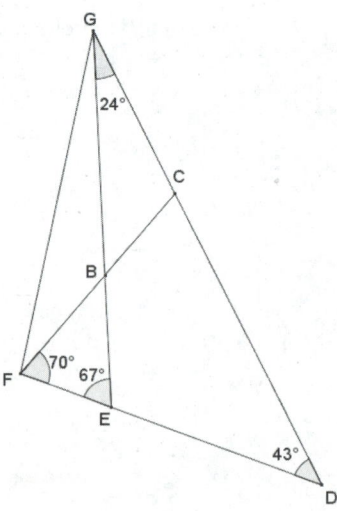

Lesson 20:   Cyclic Quadrilaterals

EUREKA
MATH™

8. In the diagram below, quadrilateral $JKLM$ is cyclic. Find the value of n.

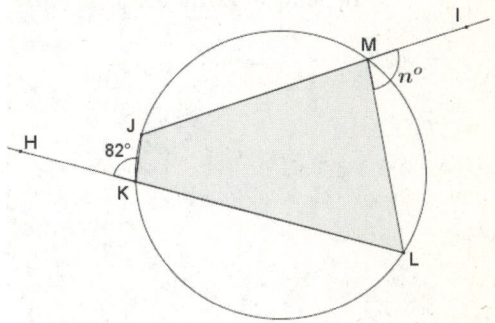

9. Do all four perpendicular bisectors of the sides of a cyclic quadrilateral pass through a common point? Explain.

10. The circles in the diagram below intersect at points A and B. If $m\angle FHG = 100°$ and $m\angle HGE = 70°$, find $m\angle GEF$ and $m\angle EFH$.

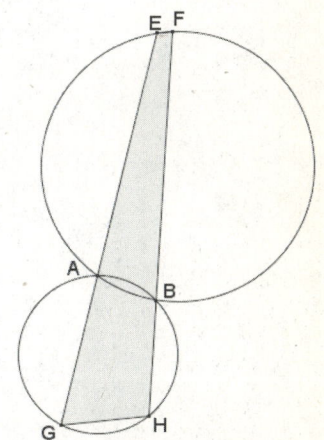

11. A quadrilateral is called *bicentric* if it is both cyclic and possesses an inscribed circle. (See diagram to the right.)

    a. What can be concluded about the opposite angles of a bicentric quadrilateral? Explain.

    b. Each side of the quadrilateral is tangent to the inscribed circle. What does this tell us about the segments contained in the sides of the quadrilateral?

    c. Based on the relationships highlighted in part (b), there are four pairs of congruent segments in the diagram. Label segments of equal length with a, b, c, and d.

    d. What do you notice about the opposite sides of the bicentric quadrilateral?

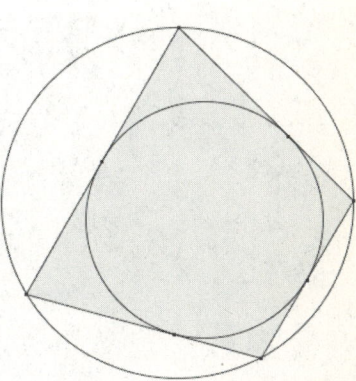

12. Quadrilateral $PSRQ$ is cyclic such that $\overline{PQ}$ is the diameter of the circle. If $\angle QRT \cong \angle QSR$, prove that $\angle PTR$ is a right angle, and show that S, X, T, and P lie on a circle.

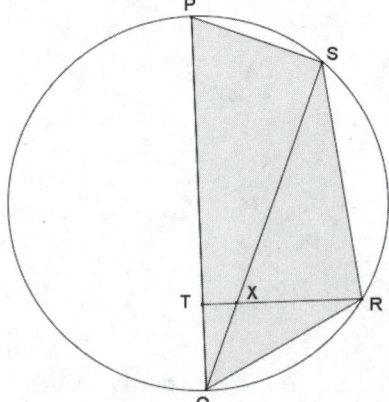

EUREKA
MATH™

Lesson 21: Ptolemy's Theorem

Classwork

Opening Exercise

Ptolemy's theorem says that for a cyclic quadrilateral $ABCD$, $AC \cdot BD = AB \cdot CD + BC \cdot AD$.

With ruler and a compass, draw an example of a cyclic quadrilateral. Label its vertices A, B, C, and D.

Draw the two diagonals $\overline{AC}$ and $\overline{BD}$.

With a ruler, test whether or not the claim that $AC \cdot BD = AB \cdot CD + BC \cdot AD$ seems to hold true.

Repeat for a second example of a cyclic quadrilateral.

Challenge: Draw a cyclic quadrilateral with one side of length zero. What shape is this cyclic quadrilateral? Does Ptolemy's claim hold true for it?

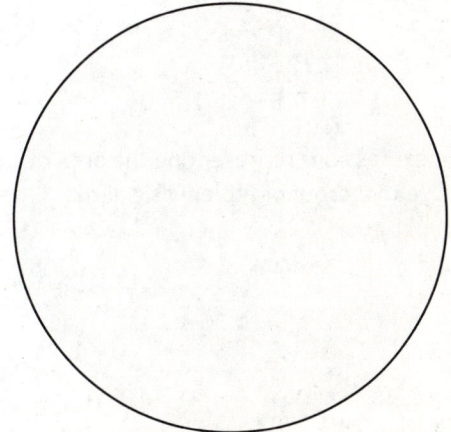

Exploratory Challenge: A Journey to Ptolemy's Theorem

The diagram shows cyclic quadrilateral $ABCD$ with diagonals $\overline{AC}$ and $\overline{BD}$ intersecting to form an acute angle with degree measure w. $AB = a$, $BC = b$, $CD = c$, and $DA = d$.

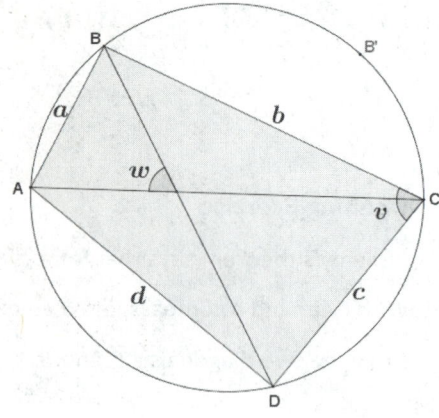

a.   From the last lesson, what is the area of quadrilateral $ABCD$ in terms of the lengths of its diagonals and the angle w? Remember this formula for later.

b.   Explain why one of the angles, $\angle BCD$ or $\angle BAD$, has a measure less than or equal to 90°.

c.   Let's assume that $\angle BCD$ in our diagram is the angle with a measure less than or equal to 90°. Call its measure v degrees. What is the area of triangle BCD in terms of b, c, and v? What is the area of triangle BAD in terms of a, d, and v? What is the area of quadrilateral $ABCD$ in terms of a, b, c, d, and v?

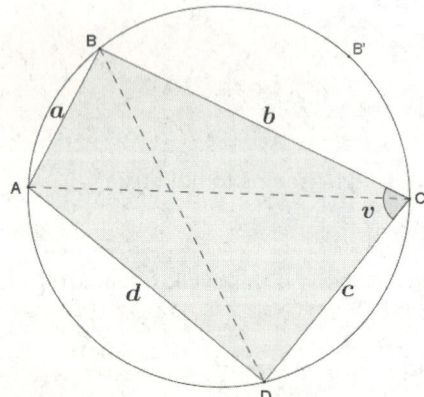

d.   We now have two different expressions representing the area of the same cyclic quadrilateral $ABCD$. Does it seem to you that we are close to a proof of Ptolemy's claim?

EUREKA
MATH™

© 2015 Great Minds. eureka-math.org
GEO-M3-SE-B2-1.3.0-10.2015

e. Trace the circle and points A, B, C, and D onto a sheet of patty paper. Reflect triangle ABC about the perpendicular bisector of diagonal $\overline{AC}$. Let A', B', and C' be the images of the points A, B, and C, respectively.

    i.   What does the reflection do with points A and C?

    ii. Is it correct to draw B' as on the circle? Explain why or why not.

    iii. Explain why quadrilateral $AB'CD$ has the same area as quadrilateral $ABCD$.

f. The diagram shows angles having degree measures u, w, x, y, and z.

    Find and label any other angles having degree measures u, w, x, y, or z, and justify your answers.

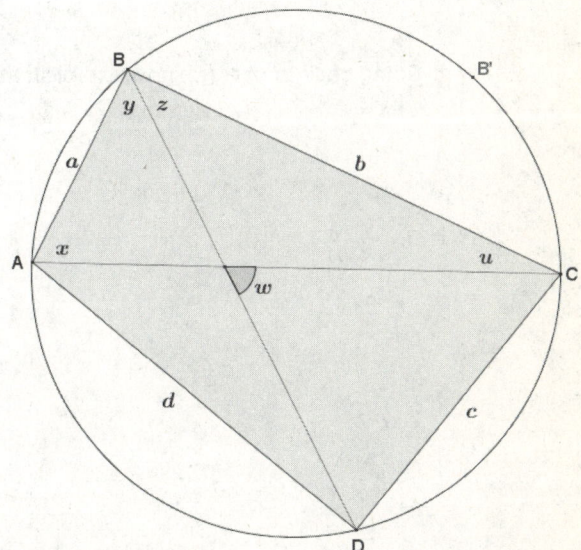

g.   Explain why $w = u + z$ in your diagram from part (f).

h.   Identify angles of measures $u, x, y, z,$ and w in your diagram of the cyclic quadrilateral $AB'CD$ from part (e).

i.   Write a formula for the area of triangle $B'AD$ in terms of $b, d,$ and w. Write a formula for the area of triangle $B'CD$ in terms of $a, c,$ and w.

j.   Based on the results of part (i), write a formula for the area of cyclic quadrilateral $ABCD$ in terms of $a, b, c, d,$ and w.

k.   Going back to part (a), now establish Ptolemy's theorem.

EUREKA
MATH

Lesson Summary

THEOREM:

PTOLEMY'S THEOREM: For a cyclic quadrilateral $ABCD$, $AC \cdot BD = AB \cdot CD + BC \cdot AD$.

Relevant Vocabulary

CYCLIC QUADRILATERAL: A quadrilateral with all vertices lying on a circle is known as a *cyclic quadrilateral*.

Problem Set

1.   An equilateral triangle is inscribed in a circle. If P is a point on the circle, what does Ptolemy's theorem have to say about the distances from this point to the three vertices of the triangle?

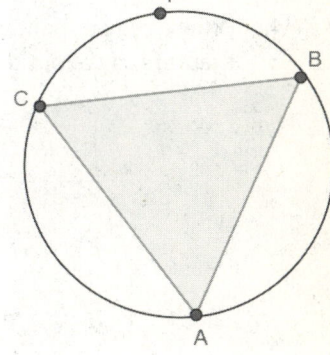

2.   Kite $ABCD$ is inscribed in a circle. The kite has an area of 108 sq. in., and the ratio of the lengths of the non-congruent adjacent sides is $3 : 1$. What is the perimeter of the kite?

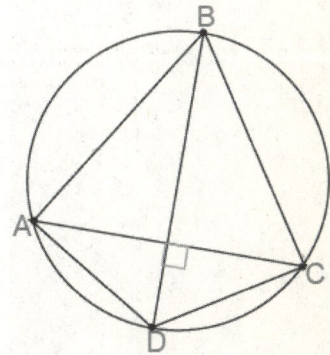

3. Draw a right triangle with leg lengths a and b and hypotenuse length c. Draw a rotated copy of the triangle such that the figures form a rectangle. What does Ptolemy have to say about this rectangle?

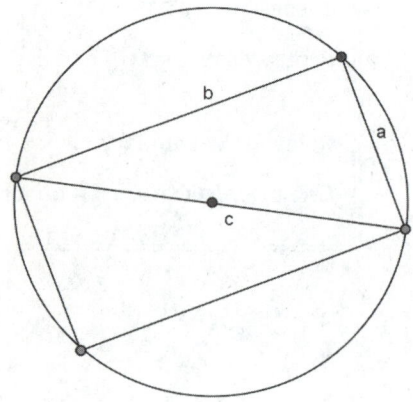

4. Draw a regular pentagon of side length 1 in a circle. Let b be the length of its diagonals. What does Ptolemy's theorem say about the quadrilateral formed by four of the vertices of the pentagon?

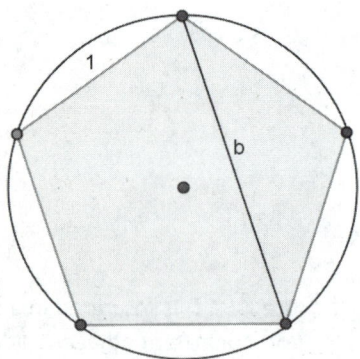

5. The area of the inscribed quadrilateral is $\sqrt{300}$ mm^2. Determine the circumference of the circle.

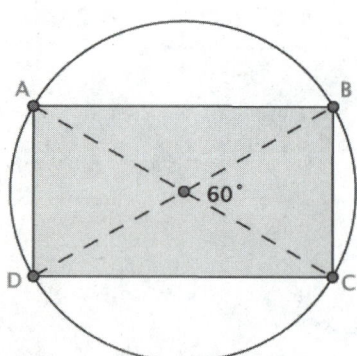

Lesson 21:   Ptolemy's Theorem

© 2015 Great Minds. eureka-math.org
GEO-M3-SE-B2-1.3.0-10.2015

6. **Extension:** Suppose x and y are two acute angles, and the circle has a diameter of 1 unit. Find a, b, c, and d in terms of x and y. Apply Ptolemy's theorem, and determine the exact value of $\sin(75)$.

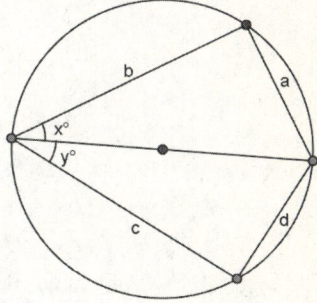

a. Explain why $\dfrac{a}{\sin(x)}$ equals the diameter of the circle.

b. If the circle has a diameter of 1, what is a?

c. Use Thales' theorem to write the side lengths in the original diagram in terms of x and y.

d. If one diagonal of the cyclic quadrilateral is 1, what is the other?

e. What does Ptolemy's theorem give?

f. Using the result from part (e), determine the exact value of $\sin(75)$.

This page intentionally left blank